i3-MARKET

Concepts and Design Innovations
Addressing the Digital Transformation
of Data Spaces and Marketplaces

i3-MARKET Book Series - Part I: A Vision to the Future of Data-Driven Economy

RIVER PUBLISHERS SERIES IN COMPUTING AND INFORMATION SCIENCE AND TECHNOLOGY

Series Editors:

K.C. CHEN
National Taiwan University, Taipei, Taiwan
University of South Florida, USA

SANDEEP SHUKLA
Virginia Tech, USA
Indian Institute of Technology Kanpur, India

The "River Publishers Series in Computing and Information Science and Technology" covers research which ushers the 21st Century into an Internet and multimedia era. Networking suggests transportation of such multimedia contents among nodes in communication and/or computer networks, to facilitate the ultimate Internet.

Theory, technologies, protocols and standards, applications/services, practice and implementation of wired/wireless networking are all within the scope of this series. Based on network and communication science, we further extend the scope for 21st Century life through the knowledge in machine learning, embedded systems, cognitive science, pattern recognition, quantum/biological/molecular computation and information processing, user behaviors and interface, and applications across healthcare and society.

Books published in the series include research monographs, edited volumes, handbooks and textbooks. The books provide professionals, researchers, educators, and advanced students in the field with an invaluable insight into the latest research and developments.

Topics included in the series are as follows:-

- Artificial intelligence
- Cognitive Science and Brian Science
- Communication/Computer Networking Technologies and Applications
- Computation and Information Processing
- Computer Architectures
- Computer networks
- Computer Science
- Embedded Systems
- Evolutionary computation
- Information Modelling
- Information Theory
- Machine Intelligence
- Neural computing and machine learning
- Parallel and Distributed Systems
- Programming Languages
- Reconfigurable Computing
- Research Informatics
- Soft computing techniques
- Software Development
- Software Engineering
- Software Maintenance

For a list of other books in this series, visit www.riverpublishers.com

i3-MARKET

Concepts and Design Innovations Addressing the Digital Transformation of Data Spaces and Marketplaces

i3-MARKET Book Series - Part I: A Vision to the Future of Data-Driven Economy

Editors

Martín Serrano

Achille Zappa

Waheed Ashraf

Pedro Maló

Márcio Mateus

Edgar Fries

Iván Martínez

Alessandro Amicone

Justina Bieliauskaite

Marina Cugurra

Routledge
Taylor & Francis Group

LONDON AND NEW YORK

Published 2024 by River Publishers
River Publishers
Alsbjergvej 10, 9260 Gistrup, Denmark
www.riverpublishers.com

Distributed exclusively by Routledge
4 Park Square, Milton Park, Abingdon, Oxon OX14 4RN
605 Third Avenue, New York, NY 10017, USA

Concepts and Design Innovations Addressing the Digital Transformation of Data Spaces and Marketplaces / by Martín Serrano, Achille Zappa, Waheed Ashraf, Pedro Maló, Márcio Mateus, Edgar Fries, Iván Martínez, Alessandro Amicone, Justina Bieliauskaite, Marina Cugurra.

ISBN: 978-87-7004-169-0 (hardback)
978-10-4009-093-0 (online)
978-10-0349-834-6 (master ebook)

DOI: 10.1201/9788770041690

Contents

i3-MARKET

Preface

Data is the oil in today's global economy. The vision in the i3-MARKET book series is that the fast-growing data marketplaces sector will mature, with a large number of data-driven opportunities for commercialization and activating new innovation channels for the data.

A new data-as-a-service paradigm where the data can be traded and commercialized securely and transparently and with total liberty at the local and global scale directly from the data producer is necessary. This new paradigm is the result of an evolution process where data producers are more active owners of the collected data while at the same time catapulting disruptive data-centric applications and services. i3-MARKET takes a step forward and provides support tools for this maturity vision/process.

i3-MARKET is a fully open source backplane platform that can be used as a set of support tools or a standalone platform implementation of data economy support services. i3-MARKET is the result of shared perspectives from a representative global group of experts, providing a common vision in data economy and identifying impacts and business opportunities in the different areas where data is produced.

Data economy is commonly referring to the diversity in the use of data to provide social benefits and have a direct impact in people's life. From a technological point of view, data economy implies technological services to underpin the delivery of data applications that bring value and address the diverse demands on selling, buying, and trading data assets. The demand and the supply side in the data is increasing exponentially, and it is being demonstrated that the value that the data has today is as relevant as any other tangible and intangible assets in the global economy.

This publication is supported with EU research funds under grant agreement i3-MARKET-871754. Intelligent, Interoperable, Integrative and deployable open source MARKETplace with trusted and secure software tools for incentivising the industry data economy and the Science Foundation Ireland research funds under grant agreement SFI/12/RC/2289_P2. Insight SFI Research Centre for Data Analytics. The European Commission and the SFI support for the production of this publication does not constitute an endorsement of the contents, which reflect the views only of the authors, and the Commission, the SFI or its authors cannot be held responsible for any use which may be made of the information contained therein.

Dr. J. Martin Serrano O.
i3-MARKET Scientific Manager and Data Scientist
Adjunct Lecturer and Senior SFI Research Fellow at University of Galway
Data Science Institute - Insight SFI Research Centre for Data Analytics
Unit Head of Internet of Things, Stream Processing and Intelligent Systems
Research Group
University of Galway, www.universityofgalway.ie | Ollscoil na Gaillimh
<jamiemartin.serranoorozco@universityofgalway.ie>
<martin.serrano@insight-centre.org>
<martin.serrano@nuigalway.ie>

i3-MARKET

Who Should Read this Book?

General Public and Students

This Book is a unique opportunity for understanding the future of data spaces and marketplace assets, their services, and their ability to identify different methodology indicators and the data-driven economy from a human-centric perspective supports the digital transformation.

Entrepreneurs and SMEs

This Book is a unique opportunity for understanding the most updated software tools to innovate, increase opportunities, and increase the power of innovation into small and entrepreneurs to meet its full potential promoting participation across the data economy values and evolution of society towards a single digital strategy.

Technical Experts and Software Developers

This book is a guide for technolgy experts and open source enthusiast that includes the most recent experiences in Europe towards innovating software technology for the financial and banking sectors.

Data Spaces & Data Markeplaces Policy Makers

This Book represent a unique offering for non-technical experts but that participates in the data economy process and the core data economy services to enable the sharing of innovation and new services across data spaces and marketplaces such as policy makers and standardisation organisatiosna and groups.

What is Addressed in the i3-MARKET Book Series?

"Concepts and Design Innovations for the Digital Transformation of Data Spaces and Data Marketplaces"

In the first part of the i3-MARKET book series, we begin by discussing the principles of the modern data economy that lead to make the society more aware about the value of the data that is produced everyday by themselves but also in a collective manner, i.e., in an industrial manufacturing plant, a smart city full of sensors generating data about the behaviours of the city and their inhabitants, and/or the wellbeing and healthcare levels of a region or specific locations. Data business is one of the most disruptive areas in today's global economy, particularly with the value that large corporates have embedded in their solutions and products as a result of the use of data from every individual.

"Systems and Implemented Technologies for Data-driven Innovation, Addressing Data Spaces and Marketplaces Semantic Interoperability Needs"

In the second i3-MARKET series book, we start reviewing the basic technological principles and software best practices and standards for implementing and deploying data spaces and data marketplaces. The book provides a definition for data-driven society as: *The process to transform data production into data economy for the people using the emerging technologies and scientific advances in data science to underpin the delivery of data economic models and services.* This book further discusses why data spaces and data marketplaces are the focus in today's data-driven society as the trend to

xiii

rapidly transforming the data perception in every aspect of our activities. In this book, technology assets that are designed and implemented following the i3-MARKET backplane reference implementation (WebRI) that uses open data, big data, IoT, and AI design principles are introduced. Moreover, the series of software assets grouped as sub-systems and composed by software artefacts are included and explained in full. Further, we describe i3-MARKET backplane tools and how these can be used for supporting marketplaces and its components including details of available data assets. Next, we provide a description of solutions developed in i3-MARKET as an overview of the potential for being the reference open source solution to improve data economy across different data marketplaces.

"Technical Innovation, Solving the Data Spaces and Marketplaces Interoperability Problems for the Global Data-driven Economy"

In the third i3-MARKET series book, we focus on including the best practices and simplest software methods and mechanisms that allow the i3-MARKET backplane reference implementation to be instantiated, tested, and validated even before the technical experts and developers community decide to integrate the i3-MARKET as a reference implementation or adopted open source software tools. In this book, the purpose of offering a guide book for technical experts and developers is addressed. This book addresses the so-called industrial deployment or pilots that need to have a clear understanding of the technological components and also the software infrastructures, thus it is important to provide the easy-to-follow steps to avoid overwhelm the deployment process.

i3-MARKET has three industrial pilots defined in terms of data resources used to deploy data-driven applications that use the most of the i3-MARKET backplane services and functionalities. The different software technologies developed, including the use of open source frameworks, within the context of the i3-MARKET are considered as a bill of software artefacts of the resources needed to perform demonstrators, proof of concepts, and prototype solutions. The i3-MARKET handbook provided can actually be used as input for configurators and developers to set up and pre-test testbeds, and, therefore, it is extremely valuable to organizations to be used properly.

i3-MARKET

What is Covered in this i3-MARKET Part I Book?

"Concepts and Design Innovations addressing the Digital Transformation of Data Spaces and Marketplaces"

In this first part of the i3-MARKET Book series we begin by discussing the principles of the modern data economy that lead to make the society more aware about the value of the data that is produced everyday by themselves but also in a collective manner, i.e. in an industrial manufacturing plant, a smart city full of sensors generating data about the behaviours of the city and their inhabitants and/or the wellbeing and healthcare levels of a region or specific locations, etc. Data Business is one of the most disruptive areas in today's global economy, particularly with the value that large corporates have embedded in their solutions and products as result of the use of data from every individual.

i3-MARKET envision a large number of data-driven opportunities emerging for activating new channels of innovation named as Data Marketplaces where the data can be trade and commercialize transparently and with full liberty at the local and global scale while at the same time catapulting data-driven business opportunities for more disruptive datacentric services. i3-MARKET is at the same time a backplane platform that can serve as a set of support tools or a standalone implementation of marketplace services as result of a sharing vision from a representative global group of experts, providing a common vision and identifying impacts in the different sectors where data is produced.

Acknowledgements

Immense thanks to our families for their incomparable affection, jollity, and constant understanding that scientific career is not a work but a lifestyle, for encouraging us to be creative, for their enormous patience during the time away from them, invested in our scientific endeavours and responsibilities, and for their understanding about our deep love to our professional life and its consequences – we love you!

To all our friends and relatives for their comprehension when we had no time to spend with them and when we were not able to join in time because we were in a conference or attending yet another meeting and for their attention and the interest they have been showing all this time to keep our friendship alive; be sure, our sacrifices are well rewarded.

To all our colleagues, staff members, and students at our respective institutions, organizations, and companies for patiently listening with apparent attention to the descriptions and progress of our work and for the great experiences and the great time spent while working together with us and the contributions provided to culminate this book series project. In particular, thanks to the support and confidence from all people who believed this series of books would be finished in time and also to those who did not trust on it, because, thanks to them, we were more motivated to culminate the project.

To the scientific community, who is our family when we are away and working far from our loved ones, for their incomparable affection, loyalty, and constant encouragement to be creative, and for their enormous patience during the time invested in understanding, presenting, and providing feedback to new concepts and ideas – sincerely to you all, thanks a million!

This publication is supported with EU research funds under grant agreement i3-MARKET-871754. Intelligent, Interoperable, Integrative and deployable open source MARKETplace with trusted and secure software tools for incentivising the industry data economy. The European Commission's support for the production of this publication does not constitute an endorsement of the contents, which reflect the views only of the authors, and the Commission or its authors cannot be held responsible for any use which may be made of the information contained therein.

Martín Serrano on Behalf of All Authors

List of Figures

List of Tables

List of Contributors

Achille, Zappa, *NUIG, Ireland*

Alessandro, Amicone, *GFT, Italy*

Andrei, Coman, *Siemens SRL, Romania*

Andres, Ojamaa, *Guardtime, Estonia*

Angel, Cataron, *Siemens SRL, Romania*

Antonio,,Jara, *Libellium/HOPU, Spain*

Birthe, Boehm, *Siemens AG (Erlangen), Germany*

Borja, Ruiz, *Atos, Spain*

Bruno, Almeida, *UNPARALLEL, Portugal*

Bruno, Michel, *IBM, Switzerland*

Carlos Miguel, Pina Vaz Gomes, *IBM, Switzerland*

Carmen, Pereira, *Atos, Spain*

Chi, Hung Le, *NUIG, Ireland*

Deborah, Goll *Digital SME, Belgium*

Dimitris, Drakoulis, *Telesto, Greece*

Edgar, Fries, *Siemens AG (Erlangen), Germany*

Fernando, Román García, *UPC, Spain*

Filia, Filippou, *Telesto, Greece*

George, Benos, *Telesto, Greece*

German, Molina, *Libellium/HOPU, Spain*

Hoan, Quoc, *NUIG, Ireland*

Iosif, Furtuna, *Siemens SRL, Romania*

Isabelle, Landreau, *IDEMIA, France*

Ivan, Martinez, *Atos, Spain*

James, Philpot, *Digital SME, Belgium*

Jean Loup, Depinay, *IDEMIA, France*

Joao, Oliveira, *UNPARALLEL, Portugal*

Jose, Luis Muñoz Tapia, *UPC, Spain*

Juan Eleazar, Escudero, *Libellium/HOPU, Spain*

Juan, Hernández Serrano, *UPC, Spain*

Juan , Salmerón, *UPC, Spain*

Justina, Bieliauskaite *Digital SME, Belgium*

Kaarel, Hanson, *Guardtime, Estonia*

Lauren, Del Giudice, *IDEMIA, France*

Luca, Marangoni, *GFT, Italy*

Lucas, Asmelash, *Digital SME, Belgium*

Lukas, Zimmerli, *IBM, Switzerland*

Márcio, Mateus, *UNPARALLEL, Portugal*

Marc, Catrisse, *UPC, Spain*

Mari, Paz Linares, *UPC, Spain*

Maria, Angeles Sanguino Gonzalez, *Atos, Spain*

Maria, Smyth, *NUIG, Ireland*

Marina, Cugurra, *ETA Consulting*

Marquart, Franz, *Siemens AG (Erlangen), Germany*

Martin, Serrano, *NUIG, Ireland*

Mirza, Fardeen Baig, *NUIG, Ireland*

Oxana, Matruglio, *Siemens AG (Erlangen), Germany*

Pascal, Duville, *IDEMIA, France*

Pedro, Ferreira, *UNPARALLEL, Portugal*

Pedro, Malo, *UNPARALLEL, Portugal*

Philippe, Hercelin, *IDEMIA, France*

Qaiser, Mehmood, *NUIG, Ireland*

Rafael, Genés, *UPC, Spain*

Raul, Santos, *Atos, Spain*

Rishabh, Chandaliya, *NUIG, Ireland*

Rupert, Gobber, *GFT, Italy*

Stefanie, Wolf, *Siemens AG (Erlangen), Germany*

Stratos, Baloutsos,*AUEB,Greece*

Susanne, Stahnke, *Siemens AG (Erlangen), Germany*

Tanel, Ojalill, *Guardtime, Estonia*

Timoleon, Farmakis, *AUEB, Greece*

Tomas, Pariente Lobo, *Atos, Spain*

Toufik, Ailane, *Siemens AG (Erlangen), Germany*

Víctor, Diví, *UPC, Spain*

Vasiliki, Koniakou, *AUEB, Greece*

Yvonne, Kovacs, *Siemens SRL, Romania*

List of Abbreviations

AI	Artificial intelligence
API	Application program interface
APP	Mobile application/web application
CA	Certificate authority
CSMT	Compact sparse merkle tree
DB	Data base
DCAT	Data catalog vocabulary
DID	Decentralized identifier
DLT	Distributed ledger technology
DSA	Data sharing agreement
ECDSA	Elliptic curve digital signature algorithm
HMAC	Hash-based message authentication code
IAM	Identity and access management
IDM	Identity management
IoT	Internet of things
IRI	Information reuse and integration
JWT	JSON web token
KOS	Knowledge organization system
NAL	Nexus authorization logic
O-CASUS	Ontology for control, access, save, use and security
OIDC	OpenID connect
OSS	Open source software
PAV	Privacy, anonymity, and verifiability
PDU	Protocol data unit
PoO	Proof of origin

PoP	Proof of publication
PoR	Proof of reception
QoS	Quality of Service
RP	Relying party
RSA	Rivest-Shamir-Adleman cryptosystem
SDA	Secure data access
SDK	Software development kit
SKOS	Simple knowledge organization system
SLA	Service level agreement
SLS	Service level specification
SME	Small and medium-sized enterprises
SQL	Structured query language
SSI	Self-sovereign identity
TLS	Transport layer security
URI	Uniform resource identifier
VC	Verifiable credentials
VDI	Verifiable database integrity
VoID	Vocabulary of interlinked datasets

1

i3-MARKET Overview

The i3-MARKET Project (www.i3-market.eu) provides solutions in the form of software artifacts implemented as open-source tools and the reference pilots deployed as industrial applications, both developed addressing the growing demand for a European Single Digital Data Market and aiming to incentivise the growing demand for a new paradigm in Data Economy.

A new paradigm in Data Economy that promotes economic growth will only be possible by creating innovative solutions that can be implemented as industrial deployments, which focuses on demonstrating that the value of the data can be distributed to all stakeholders participating in the data lifecycle. The i3-MARKET solutions aim at providing technologies for trustworthy (secure and reliable), data-driven collaboration, and federation of existing and new future marketplace platforms, with special attention to industrial data. The i3-MARKET architecture is designed to enable secure and privacy-preserving data sharing across data spaces and marketplaces, through the deployment of a backplane across operational data marketplaces.

In i3-MARKET, we are not trying to create another new marketplace, but we are implementing the backplane solutions that allow other data marketplace and data space to expand their market, facilitate the registration and discovery of data assets, facilitate the trading and sharing of data assets among providers, consumers, and owners, and provide tools to add functionalities they lack for better data sharing and trading processes.

The i3-MARKET project has built a blueprint open-source software architecture called "i3-MARKET Backplane" (www.open-source.i3-MARKET.eu) that addresses the growing demand for connecting multiple data spaces and marketplaces in a secure and federated manner.

i3-MARKET consortium is contributing with the developed software tools to build the European data market economy by innovating marketplace platforms, demonstrating with three industrial reference implementations (pilots) that a decentralized data economy and more fair growth are possible.

1

2

Architecture Overview Specification

The overall architecture defines all required components and subsystems, their basic functionality and behaviour, as well as their interfaces and interaction patterns in accordance with the user stories and the requirements specified in the project. The detailed specification of the i3-MARKET components and interfaces are reported in the chapters below.

In particular, the high-level architecture covers:

a) the i3-MARKET Backplane solutions with its core functionalities;
b) the interaction of existing data spaces and marketplaces with the i3-MARKET Backplane and each other (for secure data access) based on open interfaces;
c) the engagement of data providers, consumers, owners via smart wallets and applications, and the interactions with the i3-MARKET Backplane for the sake of privacy preservation and access control to their personal or industrial data assets.

2.1 Architecture

We describe the architecture in the 4 + 1 architectural view model. This is a standard model, commonly used for documenting software architectures.

The complete architecture is available on the i3-MARKET Wiki pages. It is available to all partners to be viewed or modified. The drawing tool used is either Gliffy or Draw.io.

2.1.1 The 4 + 1 architectural view

The *4 + 1 architectural view model* contains different views [1] as depicted in Figure 2.1.

The 4 + 1 architectural view model was adapted to fit our purposes. To make sure the different interfaces and the communication between our proposed system and the external systems are analysed, the so-called context view is added to the view model.

Table 2.1 describes the different views of the adapted 4 + 1 architectural view model.

3

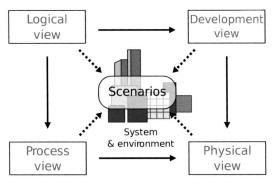

Figure 2.1 4 + 1 Architectural view model.

Table 2.1 Views of the adapted 4 + 1 architectural view model.

Architectural view	Description	Diagrams to use
Context view	• System as a blackbox • Interfaces and communication between blackbox and external systems	Context diagrams
Logical view	• Functionality that the system provides to end-users	Class and state diagrams
Process view	• Dynamic aspects of the system • System processes • Runtime behaviour of the system	Sequence, communication, and activity diagrams
Development/ implementation view	• System from a programmer's perspective • Software management	Component diagrams
Physical/deployment view	• System from a system engineer's point of view • Topology of software components on the physical layer (and their communication)	Deployment diagrams
Scenarios/use case view	• Sequence of interactions between objects and between processes • To identify architectural elements and to illustrate and validate the architecture design. • Starting point for tests	Use case diagrams

2.2 Context View

The context view shows a system as a whole and its interfaces to external factors. System context diagrams represent all external entities that may interact with a system; such a diagram pictures the system at the centre, with no details of its interior structure, surrounded by all its interacting systems, environments, and activities. The objective of the system context diagram is to focus attention on external factors and events that should be considered in developing a complete set of system requirements and constraints.

System context views are used early in a project to get agreement on the scope of the system. Context diagrams are typically included in a requirements document. These diagrams must be agreed on by all project stakeholders and thus should be written in plain language so that the stakeholders can understand items within the document.

The so-called system of interest, the i3-MARKET Backplane, is the centre of this diagram but is considered as a grey-box, showing only little internal details of the system.

The focus of this view is the external actors that have interfaces to systems. In this case, the external actors are the three pilots of the i3-MARKET project:

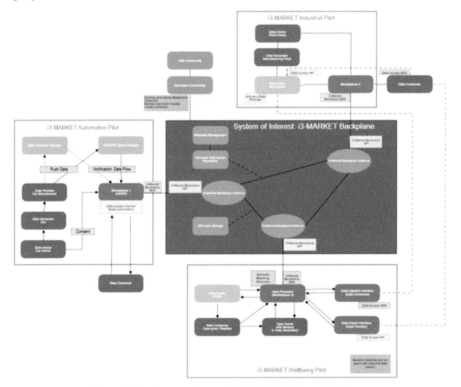

Figure 2.2   Context view with i3-MARKET as a blackbox.

- i3-MARKET automotive pilot
- i3-MARKET wellbeing pilot
- i3-MARKET industrial pilot

Figure 2.2 shows how data consumer and data provider exchange data via the data access API. The marketplace interacts with the i3-MARKET Backplane via an SDK. Both the API and the SDK were developed in the project.

The three pilot boxes also show some internal elements of the pilots. Each pilot has their own internal structure, but they share the same interface to the i3-MARKET Backplane. This enables seamless data exchange between all marketplaces.

2.3 Logical View

The logical view represented in Figure 2.3 shows the functionality that the system provides to end-users.

The objective of the logical view is twofold. On one hand, this view shows the i3-MARKET system (green box) and the link between the stakeholders and the marketplaces. On the other hand, the logical view pursues showing the internal decomposition of i3-MARKET system into the logical subsystems and components, which implement the i3-MARKET Backplane API and secure data access API (SDA API).

In general terms, i3-MARKET supports actors with the i3-MARKET Backplane functionality by means of the two following main entry points:

- The *Backplane API* and *SDA API* (depicted as green lines in the picture), or in other words, the direct access to the i3-MARKET Backplane. These two APIs enable access to all integrated building blocks. This is the use case of these actors which follow a more *ad-hoc* integration with i3-MARKET.
- The *i3-MARKET SDK (i3M SDK)* (depicted as pink boxes in the picture), to support the end-users' developers with the integration of Backplane API and SDA API. This product is intended for these actors that pursue a more "assisted" support.

Regarding the link with the stakeholder and marketplaces, in the case of the data marketplace actors, i3-MARKET assists them with a full version of the Backplane API and the i3M SDK (Backplane module), which gives support for interacting with the Backplane API.

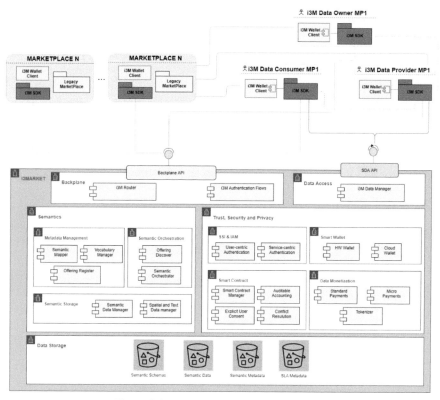

Figure 2.3 Logical view with i3-MARKET.

In the case of the data owners, data providers, and data consumers, the normal operating mode is the access to i3-MARKET Backplane through their own data marketplace. However, for some particular data marketplace cases, data owners, data providers, and data consumers will have the possibility to directly interface with i3-MARKET system through the available SDKs and APIs. More specifically, i3-MARKET will allow direct communication with the stakeholder by means of the following components:

- Data owner, through the i3M SDK (Backplane module), which gives support for interacting with the Backplane API (light green lines in the picture).
- Data provider, through the i3M SDK-Backplane module which gives support for interacting with the Backplane API (light green lines in the picture) and the i3M SDK-secure data access API, which gives support

for interacting with the secure data access API (dark green lines in the picture).
- Data consumer, through the i3M SDK-Backplane module, which gives support for interacting with the Backplane API and the i3M SDK-secure data access API, which gives support for interacting with the secure data access API.

In order to guarantee the authentication mechanisms proposed by i3-MARKET, a Wallet Client should be installed into the end-user side in order to store the user private keys.

2.4 i3-MARKET Microservice-based Architecture

i3-MARKET Backplane is mostly a set of semi-independent subsystems with self-contained functionalities such as the identity and access management system, the semantic engine subsystem, data access subsystem, etc. Most of these subsystems have broken down their functionalities into atomic and loosely coupled sub-components exposing their functionality through a REST API, which yields a microservice-based nature to the i3-MARKET system.

This microservice-based architecture brings i3-MARKET a set of very well-known benefits such as:

- facilitating the communication between the components in a system;
- have been independently developed and deployed into a more efficient management;
- facilitating the identification of dependencies between the components;
- modular architecture allows each application to use only those function-alities that are needed;
- helps to manage the complexity of the overall system.

Figure 2.4 shows a detailed landscape of the current set of microser-vices (cubes), APIs (little yellow rectangles), components (blue rectangles), and storages (white rectangles) on i3-MARKET. Each arrow in the picture denotes a dependency between the subject and object involved in the arrow. Finally, for linking each of the service/microservices/library depicted in the diagram with the component diagrams in section development view "develop-ment view", we have categorized each service/microservice/library according to the system (green dashed boxes) and subsystems (brown dashed boxes) they belong. Finally, remark that the RPC distributed ledger is one and single instance, but it has been put as several instances for picture legibility.

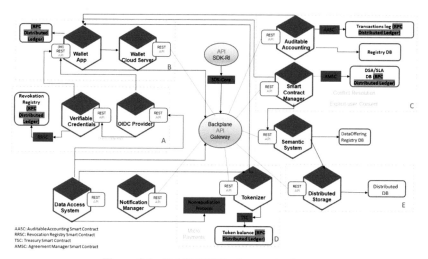

Figure 2.4 i3-MARKET microservice layout.

Figure 2.5 shows the identified dependencies between i3-MARKET components:

- SDK system: For a more grounded view of this subsystem, refer to Chapter 17 on i3-MARKET SDK and marketplace reference implementation.

  ○ *SDK-core* libraries for making easier the development of applications that make use of the Backplane API. It interfaces with the Backplane gateway.
  ○ *SDK-RI* future common pilots-driven complex workflows based on the Backplane services. It interfaces with the SDK-core library.

- Trust, security, and privacy system:

  ○ SSI and IAM subsystem (label A). For a more grounded picture of this subsystem, see "SSI and IAM subsystem" in Chapter 3 on "i3-MARKET identity and access management".

    • "User-centric authentication" component, responsible of providing the management of self-sovereign identity based on DID and VC and the compatibility with OIDC standard. Microservices:

      • *Verifiable Credential* microservice, which provides DID, Verifiable Credential management, and compatibility with OIDC standard. It interfaces with the wallet because the Verifiable Credential assumes that the user created and controls with his crypto wallet their identities and with the RPC ledger storage for updating the revocation of credential.

- "Service-centric authentication" component, responsible for providing authentication and authorization of users and client with standard OIDC/OAuth flows, integrating the user-centric authentication component. Microservices:
  - *OIDC provider* microservice, which implements the OIDC compatibility (based on Verifiable Credential). It interfaces with the Verifiable Credential for allowing the token creation based on the Verifiable Credentials and with the wallet for sending the credentials.

○ Smart Wallet subsystem (label B). For a more grounded picture of this subsystem, see "Smart Wallet subsystem" in Chapters 4 and 5.

- *Wallet APP* for storing user private keys. It interfaces on the RPC ledger storage.

○ Smart contract subsystem (label C). For a more grounded picture of this subsystem, see "smart contract subsystem" in Chapter 4 and Chapter 7 in Book Series Part II.

- *Smart contract manager* component/microservice responsible for providing a gateway to access the smart contracts. It was conceived mainly for managing the SLA and DSA smart contract (business smart contracts), and the extension of its purpose for other smart contracts is still under discussion. It interfaces with the RPC ledger storage for storing the data sharing agreement object and the semantic engine for creating data purchase.
- *Auditable accounting* component/microservice component responsible for logging and auditing interactions between components and recording the registries in the blockchain. It interfaces with the RPC ledger storage for registering auditable data and in the future might be connected with the distributed storage for storing proofs.

○ Data monetization subsystem (label D). For a more grounded picture of this subsystem, see "data monetization subsystem" in Text on "i3-MARKET Crypto token and data monetization".

○ *Non-repudiation protocol library.* It interfaces with the Backplane API for interacting with auditable accounting.

- Semantic system:

○ *Semantic system* service responsible for managing the offerings/discovery and semantic data model in the i3-MARKET. It interfaces with contract manager managing contractual parameters, and it depends on the registry DB store and the distributed DB service's API. It might interact with ledger for Verifiable Credentials and DID IDs. And it interacts with the notification manager service for reporting new data offerings. For a more grounded picture of this component, see Chapter 4 "i3-MARKET Semantic Models" and Chapter 9 in Book Series Part II "i3-MARKET Semantic Model Repository and Community."

- Data access system:

○ *Data access* service in charge of providing the means for allowing the transfer of data between data provider and the data consumer. It interacts with the non-repudiation protocol library and Backplane (API) for enforcing smart contract. For a more grounded picture of this system, see Chapter 6 on data access & transfer – system.

- Storage system:

○ Distributed storage subsystem (label D):

1. *Distributed storage* component/microservice responsible of storing i3-MARKET offerings index. Other uses of the distributed storage are still under discussion. Its interaction with the RPC ledger storage for storing proof for reliability of data is still under discussion.

- *Backplane gateway* component responsible for providing a gateway for all the internal services conforming to the Backplane. This gateway is the single-entry point for all clients. For more details about this component, see Chapter 3 in Book Series Part III Backplane API gateway. It depends on the availability of all the services masked behind it.

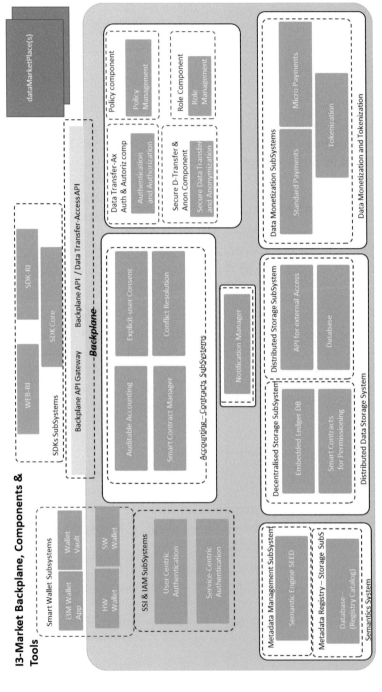

Figure 2.5 High-level view of the i3-MARKET Backplane architecture and blocks for systems and artefacts.

2.5 i3-MARKET Core Functionality

i3-MARKET provides a set of core components in charge of providing the following capabilities:

1. **Authentication-identity/authorization-access:** i3-MARKET should allow users to authenticate themselves and get authorization to access the blockchain, secure and centralized data storage, and secured services on i3-MARKET.

2. **User management:** i3-MARKET should allow data providers and consumers to register, update, or delete from the system. In particular, i3-MARKET assists the data marketplaces and stakeholders with following functionalities: identity creation, user registration, identity update, and user deletion (which ensures that his identity on the blockchain cannot be mapped to his real identity anymore. → automatically consent termination/cancellation).

3. **Offering registration:** i3-MARKET must provide mechanisms for data providers to publish their datasets on i3-MARKET. i3-MARKET provides semantic data models to describe data offerings and data subscriptions/demands.

4. **Offering discovery:** i3-MARKET must provide mechanisms to allow data consumers to perform data queries based on the provided semantic models. It will have two variants: discover and retrieve locally or discover and retrieve in federated i3-MARKET data marketplaces network.

5. **Data subscription:** i3-MARKET must provide mechanisms to allow data consumers to express the intention of buying data and to request an SLA/SLS between the data consumer and a data provider, after a match was encountered.

6. **Consent:** i3-MARKET has to provide the mechanisms in order for the data owner to consent access for enabling the trading of data assets across domains and stakeholder boundaries, without the need for developers of an application (data consumer) to learn about the meaning of the data from the data provider or through manual analysis or experimentation with the data.

   a. Explicit consent: i3-MARKET must provide mechanisms to allow the data owner to give his consent before his data is transferred (data owner). When a user is deleted, all the data and metadata related to the user should be removed from any platform.

    b. Consent termination/cancellation: i3-MARKET must provide the mechanisms for ending the commercialization between involved parties on a smart contract. i3-MARKET provides the mechanisms to end running smart contracts at any time.

7. **Contracting:** i3-MARKET has to provide mechanisms that allow to complete data sharing agreements (SLA/SLS) between the data provider and the data consumer. The smart contracts are then generated from the SLA/SLS, which the participants agreed upon (data provider and data consumer).

8. **Data access:** i3-MARKET provides a data access API enabling an authorization of the data provider and the data consumer to allow a secure data transfer (peer-to-peer or i3-MARKET-channel subscription). The data access API provides a mechanism to monitor the data transfer and is tightly coupled with the signed smart contracts. This functionality was broken down into the following modules:

    i. Authentication and authorization
    ii. Data transfer transparency
    iii. Data management
    iv. Secure data transfer and anonymization

On the other hand, i3-MARKET supports the following types of data transfer:

    a. On-demand → Data stream (see common vocabulary below).
    b. Subscription → Data batch transfer (see common vocabulary below).

9. **Data monetization/payment:** i3-MARKET provides functionality for data monetization, which aligns based on the pricing model defined in the offering description and amount consumed.

These capabilities have been validated in the i3-MARKET basic workflows (described in the following section). In concrete:

- "User management": For all end-users (data providers and data consumers), an identity should be created in advance for getting authentication-identity and authorization-access. Therefore, a user management activity will take place as pre-requisite for starting any interaction with any i3-MARKET instance.

- "Authentication-identity/authorization-access" is used as starting point for initiating any connection with i3-MARKET instances. Therefore, the process of authentication (and authorization) can be reflected at the beginning of most of the workflows. These are:

  ○ "Registering a new offering": The workflow starts with the authentication of the data provider as described in the diagram "authentication with end-user interaction" in Chapter 4 "i3-MARKET Semantic Models".
  ○ "Purchase data", "create and manage search alerts", and "transfer operational data": The workflow starts with the authentication of the data provider and data consumer, as described in the diagram "authentication with end-user interaction" in Chapter 4 "i3-MARKET Semantic Models".

- "Offering registration", the behaviour of the offering registration capability is directly shown in the "register a new offering" workflow.
- "Offering discovery", "data subscription", and "contracting" capabilities take place in the "purchase data" workflow.
- "Data access" and "data monetization" are the most significant steps in the "transfer operational data" workflow.

2.6 i3-MARKET Basic Workflows

i3-MARKET includes the implementation of three pilots to validate the functionalities of the i3-MARKET network. Even though every pilot has its own way of how it works and its specific requirements, there are some basic workflows, which apply to all of them. Those workflows are described for the five most important scenarios.

The scenarios are:

- Generate data.
- Register new data offerings.
- Search, discover, and retrieve data offerings in local and federated registry catalogues.
- Purchase data.
- Manage notifications.
- Access and transfer operational data.

The scenarios are described in detail in i3-MARKET deliverables about "Use Cases, Requirements and Overall i3-MARKET Architecture Specifications" available at https://www.i3-market.eu/research-and-technology-library/.

For each of the scenarios (which are part of the problem space), technical workflows have been derived. These workflows represent the technical realization in the solution space. They represent the dynamic behaviour of the system when stakeholders interact with it.

2.7 Process View

According to [2], the process view "takes into account some non-functional requirements, such as performance and availability. It addresses issues of concurrency and distribution, of system's integrity, of fault-tolerance, and how the main abstractions from the logical view fit within the process architecture—on which thread of control is an operation for an object actually executed".

Following this approach, the i3-MARKET process view should show the interaction between the process and threads of the system representing, among others, non-functional characteristics, concurrency, synchronization, availability, or performance.

From a diagram point of view, Booch stated "... the static and dynamic aspects of this view are captured in the same kinds of diagrams as for the design view – i.e. class diagrams, interaction diagrams, activity diagrams and statechart diagrams, but with a focus on the active classes that represent these threads and processes" [3].

A fine-grained detail is demanded for identifying the active objects (process and threads), which are instances of active classes, and the way they communicate between each other (synchronous/asynchronous) which is needed for this view. Due to that, the i3-MARKET process view definition was accomplished between the different technical task implementations.

2.8 Development View

According to [2], the development view "focuses on the actual software module organization...The software is packaged in small chunks—program libraries, or subsystems—that can be developed by one or a small number of developers. The subsystems are organized in a hierarchy of layers, each layer providing a narrow and well-defined interface to the layers above it".

Therefore, the objective of the development view is twofold:

- Giving a system view from a programmer's perspective, which might help in the development process.
- Supporting the software management by monitoring the accomplishment of subsystems and components depicted in the diagrams.

For the process of defining the development view, i3-MARKET has followed the guidelines proposed by arc42 template [4] for architecture construction and documentation, which is summarized in the following section.

2.8.1 Approach

The i3-MARKET "development view" is documented following the "building block view" as depicted in Figure 2.6 from arc42 template. Following are the cited instructions provided by the template:

- **Content:**

The building block view shows the static decomposition of the system into building blocks (modules, components, subsystems, classes, interfaces, packages, libraries, frameworks, layers, partitions, tiers, functions, macros, operations, data structures, etc.) as well as their dependencies (relationships, associations, etc.).
This view is mandatory for every architecture documentation. In analogy to a house, this is the floor plan.

- **Motivation:**

Maintain an overview of your source code by making its structure understandable through abstraction.
This allows you to communicate with your stakeholder on an abstract level without disclosing implementation details.

- **Form:**

The building block view is a hierarchical collection of black boxes and white boxes (see Figure 2.6) and their descriptions.

Level 1 is the white box description of the overall system together with black box descriptions of all contained building blocks.

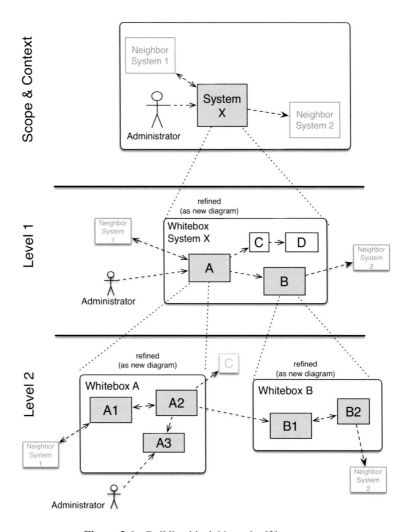

Figure 2.6   Building block hierarchy [3].

Level 2 *zooms into some building blocks of level 1. Thus, it contains the white box description of selected building blocks of level 1, together with black box descriptions of their internal building blocks.*

Level 3 *zooms into selected building blocks of level 2, and so on.*

In i3-MARKET, we have the following arc42-based templates for the documentation of the Level 1 development view.

Here you describe the decomposition of the overall system using the following white box template. It contains:

- an overview diagram;
- a motivation for the decomposition;
- black box descriptions of the contained building blocks (use a list of black box descriptions of the building blocks according to the black box template (see below). Depending on your choice of tool, this list could be sub-chapters (in text files), sub-pages (in a Wiki), or nested elements (in a modelling tool)".

3

i3-MARKET Trustworthy Design

3.1 Objectives

Today, users' identities and related data are stored in siloes on centralized servers across organizations and are vulnerable to hacking. Repetitive account creation for different applications (e.g., marketplaces) and personal information (often outdated) stored in various services are some other drawbacks of that approach. Distributed/self-sovereign identities supported by decentralized systems come as a solution to those issues and facilitate interoperability, ensuring security and compliance with the privacy regulation.

Figure 3.1 shows the overall architecture of i3-MARKET. We implemented the reference implementation for the identity and access management system of i3-MARKET Backplane based on the self-sovereign identity paradigm.

The idea behind these specifications is to provide self-sovereign identity capabilities, based on distributed identity and Verifiable Credentials concepts, maintaining the most used authentication and authorization flows and standards in this moment, to facilitate the integration of stakeholders' applications and incentivize a wide adoption.

The final implementation of the i3-MARKET IAM system is based on the selected open-source technologies for SSI (Veramo), OIDC (panva/node-oidc-provider), and standard IAM.

The user-centric authentication is provided by the Verifiable Credentials micro service and the OIDC SSI Auth micro service developed using the Veramo framework.

The choice of Veramo as SSI technology has been driven from the maturity and readiness level of the uPort technology with respect to the other state-of-the-art technologies evaluated (Hyperledger Aries, Sidetree) and for the compatibility to the blockchain chosen for i3-MARKET (Hyperledger BESU), and then, after the uPort company announced the launch of this new technology (which makes uPort deprecated), we evaluated it as a very promising framework and decided to adopt it for the final implementation.

21

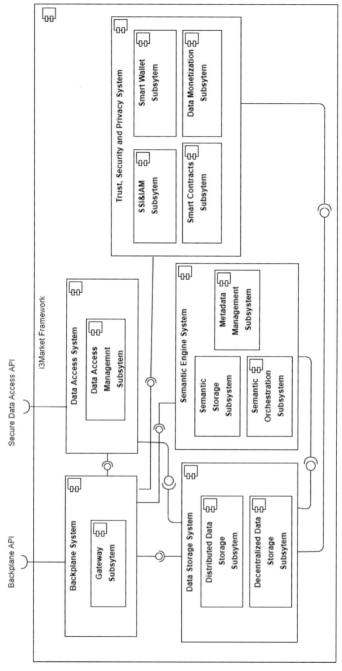

Figure 3.1   Backplane architecture.

Finally, the user-centric authentication components were integrated with the i3-MARKET Smart Wallet, implementing the pairing flow and modifying the issuing and request of Verifiable Credentials in the registration and login flows.

The following are the high-level capabilities provided by the SSI and IAM subsystems:

1. User-centric authentication: authentication of users based on the self-sovereign identity paradigm.
2. Service-centric authentication: authentication of clients and users based on a standard OIDC/OAuth identity and access management system.

3.1.1 Context

The SSI and IAM subsystem block interacts with the following three building blocks:

- Data storage system: The SSI and IAM subsystem uses the data storage system, in particular the decentralized data storage component, for recording a DID document.
- Backplane system: The SSI and IAM subsystem is used from the Backplane system for authenticating and authorizing users and clients.
- Data access system: The SSI and IAM subsystem is used from the data access system for authenticating and authorizing users and clients.

3.1.2 Building block big picture

The specific SSI and IAM component diagram is shown in Figure 3.2.

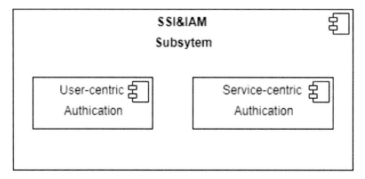

Figure 3.2 SSI and IAM components.

The SSI and IAM subsystem is in charge of providing both "user-centric authentication" and "service-centric authentication" capabilities.

Inside, we can find the following:

- A "user-centric authentication" component, responsible for providing the management of self-sovereign identity based on DID and VC and the compatibility with the OIDC standard.
- A "service-centric authentication" component, responsible for providing authentication and authorization of users and client with standard OIDC/OAuth flows, integrating the user-centric authentication component.

3.2 Technical Requirements

The list of technical requirements was not within an elicitation process to ensure the collection of not only technical needs but also service and application needs from the different stakeholders (actors) as it is described in each table from Table 3.1.

3.2.1 Actors

Table 3.1   Actors of the system.

Name	Description	Labels
Data provider	Actor who receives raw data from data owners and push it to the marketplace	Data provider
Data owner	Actor who generates the data and therefore the ultimate owner of the data. Data owners have to accept data requests to generate contracts, which leads to share the data with data consumers	Data owner
Data consumer	Consumer data shared from data owners has to create data requests through the data discovery in order for data owners to accept them. Data consumers receive only the data they want	Data consumer
Administrator	Manages the marketplace and its users	Administrator

3.2.2 User-centric authentication

The human in the loop, also known as the human-centric approach, is a design consideration that is considered in i3-MARKET. The design consideration and requirements are described in Tables 3.2 and 3.3.

3.2.2.1 Epics

Table 3.2   Epics of user-centric authentication.

Name	Description	Labels
DID management	A decentralized system that enables several key actions by three distinct entities: the controller, the relying party, and the subject. Controllers create and control DIDs, while relying parties rely on DIDs as an identifier for interactions related to the DID subject. The subject is the entity referred to by the DID, which can be anything: a person, an organization, a device, a location, or even a concept. Typically, the subject is also the controller.	Data Consumer Data Marketplace Data Provider Data Owner
Verifiable Credentials management	Verifiable Credential is a tamper-evident credential that has authorship that can be cryptographically verified though a proof. It can be used to share and prove something about the identity of a user.	Data Marketplace Data Consumer Data Provider Data Owner
OIDC client compatibility	• Staying backward compatible with existing OIDC clients and OPs that implement the OIDC specification to reach a broader community. • Adding scopes and validation rules based on VC for OIDC clients to make full use of DIDs. • Not relying on any intermediary such as a traditional centralized public or private OP while still being OIDC-compliant.	Data Marketplace

3.2.2.2 User stories

Table 3.3   User stories of user-centric authentication.

Name	Description	Labels
Create DID	As a subject, I want to create a DID so that I can manage my identity Subject: Data Consumer, Data Provider, and Data Owner	User Story Data Consumer Data Provider Data Owner
Present DID	As a user, I want to present my DID to a relying party so that I can identify myself User: Data Consumer, Data Provider, Data Owner Relying party: Data Marketplace, Data Provider	Data Consumer Data Owner Data Provider User Story
Rotate DID	As a user, I want to change the ownership of my DID so that I can maintain my identity if I change the identity provider	User Story Data Consumer Data Provider Data Owner
Delegate DID	As a user, I want to delegate my DID so that I can make other DIDs able to act on behalf of me	User Story Data Consumer Data Provider Data Owner
Recover DID	As a user, I want to recover my DID so that I can maintain my identity even if I lose my proof of control User: Data Consumer, Data Provider, Data Owner	User Story Data Consumer Data Provider Data Owner
Sign assets	As a user, I want to sign my assets so that I can demonstrate the authenticity of the asset User: Data Consumer, Data Provider, Data Owner	User Story Data Consumer Data Provider Data Owner
Verify asset signature	As a user, I want to verify asset signature so that I can authenticate the asset User: Data Consumer	User Story Data Consumer
Deactivate DID	As a user, I want to deactivate my DID so that I can delete my identity User: Data Consumer, Data Provider, Data Owner	User Story Data Consumer Data Provider Data Owner
Resolve DID	As a data marketplace, I want to resolve DID so that I can retrieve from a DID document the information to authenticate a DID subject and verify data asset signature	Data Marketplace Data Provider User Story
Authenticate DID	As a relying party, I want to authenticate DID so that I can verify the DID ownership Relying Party: Data Marketplace/Data Provider	User Story Data Marketplace Data Provider

Table 3.3 Continued.

Name	Description	Labels
Create Verifiable Credential	As a data marketplace, I want to create a Verifiable Credential so that I can provide a user an attestation of his/her role	User Story Data Marketplace
Issue Verifiable Credential	As a data marketplace, I want to issue a Verifiable Credential so that I can attest something about my users	User Story Data Marketplace
Receive Verifiable Credential	As a user, I want to receive a Verifiable Credential so that I can access the data marketplace	User Story Data Marketplace Data Provider
Store Verifiable Credential	As user, I want to store a Verifiable Credential so that I can use and keep it and use it towards any relying party	User Story Data Consumer Data Provider Data Owner
Request Verifiable Credential	As a data marketplace/data provider, I want to request Verifiable Credentials for the authenticated user so that I can give the right access to my resources	User Story Data Consumer Data Provider Data Owner
Share Verifiable Credential	As a user, I want to share a Verifiable Credential so that I can attest something towards a relying party	User Story Data Consumer Data Provider Data Owner
Verify Verifiable Credential	As a user, I want to receive a Verifiable Credential so that I can access a data marketplace	User Story Data Marketplace Data Provider
Keep track of issued Verifiable Credentials	As an issuer, I want to keep track of issued Verifiable Credentials so that I can monitor and revoke them	User Story Data Marketplace
Revoke Verifiable Credential	As an issuer, I want to revoke a Verifiable Credential so that it cannot be used	User Story Data Marketplace
OIDC authentication	As a relying party (RP), I want to authenticate users based on OIDC standards so that I do not have to change my OIDC clients RP: Data Marketplace, Data Provider	User Story Data Marketplace Data Provider

3.2.3 Service-centric authentication

The design of i3-MARKET also includes service-centric consideration; today, micro services are a trend, but this may change. Thus, the design principles are described in Tables 3.4 and 3.5.

3.2.3.1 Epics

Table 3.4   Epics of service-centric authentication.

Name	Description	Labels
Existing identity provider integration	Run a standard OpenID Connect relaying party (or OAuth2 client) on the Backplane API	Epic Data Marketplace

3.2.3.2 User stories

Table 3.5   User stories of service-centric authentication.

Name	Description	Labels
Existing identity provider authentication	As a data marketplace, I want to authenticate my users using approved external identity providers	User Story Data Marketplace

3.3 Solution Design

3.3.1 User-centric authentication

In order to provide authentication and authorization with distributed identity and Verifiable Credentials, we implemented two Node.js micro services. The Verifiable Credential micro service provides the APIs that implement the core functions to manage Verifiable Credentials, namely issuing, verifying, and revoking Verifiable Credentials, and a utility function. The OIDC SSI Auth micro service provides the API to perform the authorization code flow with PKCE using Verifiable Credentials as a proof method.

To implement the solution, we have chosen Veramo (https://veramo.io/), a framework that replaces the previous implementation of the uPort library, which is deprecated.

Both components (OIDC SSI Auth and Verifiable Credential micro service) integrate the Veramo framework and take advantage of its features to manage DID and Verifiable Credentials in Figure 3.3.

The i3-MARKET network is composed of different data marketplaces running an instance of the i3-MARKET Backplane connector. Each of them has its own OIDC SSI Auth Service and its own Verifiable Credential

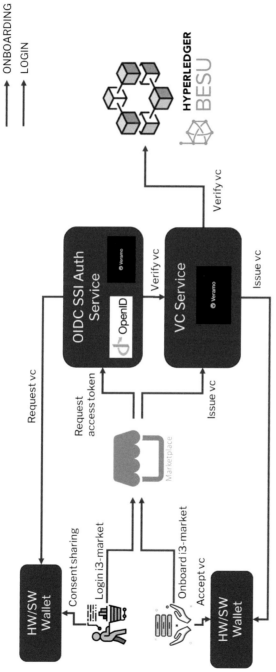

Figure 3.3 OIDC SSI Auth Service architecture.

micro service to generate, verify, and revoke Verifiable Credentials. In relation to the roles of the W3C Recommendations on verifiable credentials (https://www.w3.org/TR/vc-data-model/), the OIDC SSI Auth Service is the verifier, the Verifiable Credential micro service is the issuer (with some extra features), and the user is clearly the holder of his Verifiable Credentials. Each instance of the Verifiable Credential micro service has its own DID (https://www.w3.org/TR/did-core/) and private key used to sign Verifiable Credentials. In this way, each Verifiable Credential has as its issuer the DID of the micro service that generated it. Similarly, for revocation, only the micro service that generated a credential has the privilege to revoke it.

The user saves the Verifiable Credentials in his wallet and gives an explicit consent to share them with the OIDC SSI Auth Service when requested during the authentication phase.

The modules and detailed workflows are presented in the following subsections.

- **DID management:**

DID management is provided by Hyperledger BESU blockchain and the Veramo Ethr-DID library.

This library conforms to ERC-1056 and is intended to use Ethereum addresses as fully self-managed decentralized identifiers (DIDs).

Ethr-DID provides a scalable identity method for Ethereum addresses, which gives any Ethereum address the ability to collect on-chain and off-chain data.

This particular DID method relies on the Ethr-Did-Registry. The Ethr-DID-Registry is a smart contract that facilitates public key resolution for off-chain (and on-chain) authentication. It also facilitates key rotation, delegate assignment, and revocation to allow third-party signers on a key's behalf, as well as setting and revoking off-chain attribute data. These interactions and events are used in aggregate to form a DID document using the Ethr-Did-Resolver as shown in Figure 3.4.

DID management supports the proposed decentralized identifiers spec from the W3C Credentials Community Group.

This library has been integrated both in OIDC SSI Auth and Verifiable Credentials micro services to resolve and authenticate DID interacting with the user's wallet.

```
{
  '@context': 'https://w3id.org/did/v1',
  id: 'did:ethr:0xb9c5714089478a327f09197987f16f9e5d936e8a',
  publicKey: [{
      id: 'did:ethr:0xb9c5714089478a327f09197987f16f9e5d936e8a#owner',
      type: 'Secp256k1VerificationKey2018',
      owner: 'did:ethr:0xb9c5714089478a327f09197987f16f9e5d936e8a',
      ethereumAddress: '0xb9c5714089478a327f09197987f16f9e5d936e8a'}],
  authentication: [{
      type: 'Secp256k1SignatureAuthentication2018',
      publicKey:
  'did:ethr:0xb9c5714089478a327f09197987f16f9e5d936e8a#owner'}]
}
```

Figure 3.4 Example of a DID document resolved.

The library has been used by the Verifiable Credentials micro service to create and manage the distributed identity issuing the credentials while the distributed identities of the users must be created and managed by the user's wallet.

- **Verifiable Credential management:**

For the Verifiable Credential management, the Verifiable Credentials micro service uses the Veramo framework to generate the credentials and call the i3-MARKET Wallet API to provide the credential to the user.

- **Issue a Verifiable Credential:**

The first scenario in which a data marketplace issues a Verifiable Credential to a user is the registration process. In this scenario, the micro service has to authenticate the DID of the user and then issue for this DID a Verifiable Credential that certifies the role of the user, which can be a data consumer, a data provider, or both. The workflow for the registration process is described in Figure 3.5. The entities involved are the following:

- ○ the identity holder, which is the i3-MARKET user;
- ○ the user agent, which is also the client of the OIDC (i3-MARKET data marketplace website);
- ○ the i3-MARKET wallet, which is the wallet in which credentials are stored;
- ○ the Verifiable Credential micro service (i3-MARKET data marketplace instance);

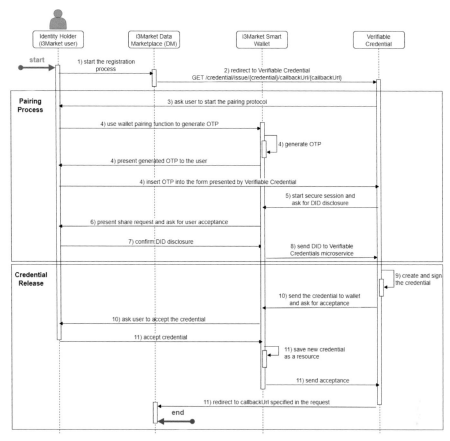

Figure 3.5   User registration flow.

  ○ the i3-MARKET data marketplace backend.

  The user registration flow is shown in Figure 3.5. At the beginning, the user enters his registration data, relating to a data consumer user or a data provider. When a user is registered as a data consumer or data provider, he is registered for all the sites of the i3-MARKET network. For this reason, these two Verifiable Credentials are issued exclusively by the i3-MARKET data marketplace entity. These data are entered on registration forms in a dedicated section of the i3-MARKET powered data marketplace.

  In order to know for which DID the Verifiable Credential should be issued, the i3-MARKET data marketplace must obtain the DID of the user. This part of the flow involves the *wallet-protocol session* API, which is specifically designed to open a secure connection ("paring") with the wallet (see deliverables in "Trust, Security and Privacy Solutions for Securing Data Marketplaces" at https://www.i3-market.eu/research-and-technology-library/ for more details), using a generated OTP, to retrieve the DID of the user. In particular, the i3-MARKET data marketplace performs a GET to the *issue* API of the Verifiable Credentials micro service passing as parameters a *callbackUrl* (which indicates the URL where to redirect the user after the issue of the credential) and the *credential* formatted as JSON encoded object.

The i3-MARKET data marketplace initiates this API call, and the Verifiable Credential micro service uses the "pairing" protocol to connect to the wallet and asks for an OTP to connect to the i3-MARKET Smart Wallet.

The user generates a new OTP, using the related wallet function and presents it to the Verifiable Credential to start a secure session (Figure 3.6). Then the Verifiable Credential sends a share request to retrieve user DID.

At this point, the user receives the disclosure request through the wallet and decides whether to accept or not to share the requested identity (DID)

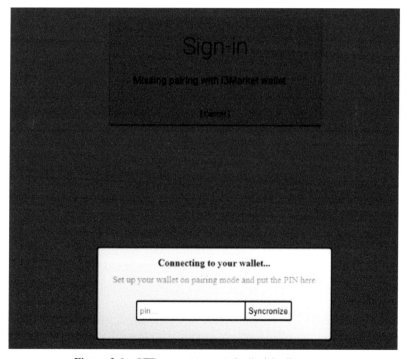

Figure 3.6 OTP request to start the "pairing" process.

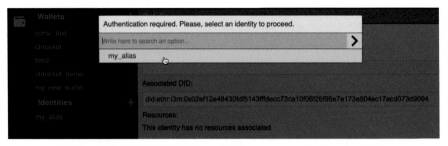

Figure 3.7   i3-MARKET Smart Wallet request to disclose the DID.

(Figure 3.7). If the user agrees to share the DID, the i3-MARKET Smart Wallet sends the following access token to the Verifiable Credentials micro service via callback. At this point, the Verifiable Credentials service decodes the access token to extract the DID of the user who has authenticated.

The second part of the flow relates to the issue of a Verifiable Credential to certify that the user is a data consumer. At a high level, issuing a Verifiable Credential involves two steps:

- Cryptographically signing the credential data.
- Sending the signed credential as a JWT (https://datatracker.ietf.org/doc /html/rfc7519) to the i3-MARKET Smart Wallet.

In order to create a Verifiable Credential, the Verifiable Credential micro service performs an internal API call to the POST/credential/issue/{DID} endpoint, communicating the user's DID just retrieved and the credential as form-data in the following format:

```
{
"data_consumer": true
}
```

The i3-MARKET data marketplace may request the issuance of credentials only relating to the registration process, i.e., data consumer and data provider. All other credentials, relating to the purchase of assets or services, can be requested by Data Providers. When the API is called, the Verifiable Credential micro service performs the Veramo *createVerifiableCredential* function, provided by the DID agent of the Veramo Core library as shown in Figure 3.8.

When the i3-MARKET Smart Wallet receives the credential, it verifies its signature. Each signed message has an "iss" attribute that contains a DID of the issuer. To resolve the public key of the message, a DID-resolver is used. The Veramo DID agent currently supports many DID methods, such as

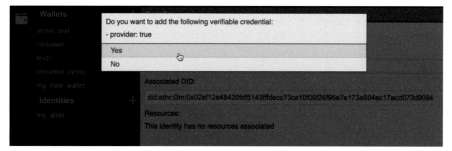

Figure 3.8 Verifiable Credential acceptance.

"did:ethr" (based on ERC-1056), "did:web" (in conjunction with blockchain-based DID, it can bootstrap trust using a web domain's existing reputation), and "did:key" (self-certifying DID method, which does not require any external utility such as blockchain). More details about supported DID methods can be found in Veramo documentation.

Being Hyperledger Besu the reference blockchain, the users' DID is "did:ethr".

After the signature verification, the wallet asks the user to accept the credential. When the user accepts the credential, it will be saved in his wallet and then be present and visible in the resources tab, which contains the list of the credentials registered in the app as shown in Figure 3.9.

At this point, the user will be redirected to the *callbackUrl*, previously specified.

Figure 3.9 Credentials list.

Once a user has saved some credentials in his wallet, he can disclose them in authentication requests, in order to certify that he holds the credentials needed to access resources or services.

- **Revoke a Verifiable Credential:**

As part of the process of working with Verifiable Credentials, it is not only necessary to issue credentials, but sometimes it is also necessary to revoke them. The ability to revoke a credential when it is no longer valid is a core function in a Verifiable Credential ecosystem. For example, suppose an i3-MARKET data provider issues a credential to access a service, and a data consumer violates the terms of use. The data provider determines that the user has violated the terms of use and, consequently, wants to suspend access to the service. In this way, the status of the Verifiable Credential needs to be changed and the next time a relying party checks the status, they will be able to see that the user is no longer valid and consequently not authorized to access the service. In order to satisfy this requirement, an API to revoke credentials has been implemented and the workflow to revoke a credential is described with Figure 3.10.

At the beginning of the flow, the data provider calls the API of the Verifiable Credentials micro service (1) communicating that a specific credential belonging to a user must be marked as revoked. The Verifiable Credential to be revoked is passed through the body parameter in the form {"JWT": "eyJhbGc ..."}.

As an implementation choice, it was decided that only those who issued a credential are allowed to revoke it. To satisfy this requirement, a check

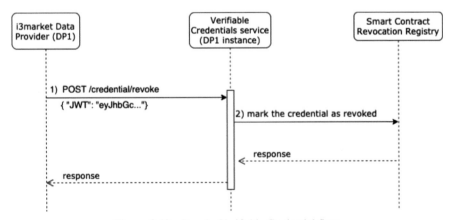

Figure 3.10   Revoke Verifiable Credential flow.

is made, if the issuer of the credential is the address of the issuer, then it proceeds; otherwise, it immediately blocks the flow. When the Verifiable Credentials micro service receives the API call, it writes the credential hash through a transaction in a smart contract named RevocationRegistry (2), in order to keep track of the action performed. Since the i3-MARKET blockchain is an Ethereum-type blockchain, the smart contract is written in solidity and its code is the following:

```
1   pragma solidity ^0.5.8;
2
3   contract RevocationRegistry {
4
5       mapping(bytes32 => mapping(address => uint)) private revocations;
6
7       function revoke(bytes32 digest) public {
8           require (revocations[digest][msg.sender] == 0);
9           revocations[digest][msg.sender] = block.number;
10          emit Revoked(msg.sender, digest);
11      }
12
13      function revoked(address issuer, bytes32 digest) public view returns (uint) {
14          return revocations[digest][issuer];
15      }
16
17      event Revoked(address issuer, bytes32 digest);
18  }
```

The smart contract RevocationRegistry provides two functions:

- a public function to revoke a credential;
- a public function to check if a credential is in the revocation list, i.e., it has been revoked.

The revoke function takes as input a string of 32 characters and writes a record associating it with the sender of the transaction, i.e., the address commits the line. In order to always have 32 characters, the credential before being marked on the smart contract is processed by a SHA-3 cryptographic hash algorithm and the 32-character digest is written on the smart contract.

The data structure of the smart contract is a private array of digest-address associations, named revocations (line 5). Whenever a credential is added to the register, it is mapped via the credential digest and the issuer of the transaction, i.e., the message sender. On that mapping, the block number is written, i.e., the transaction counter ID.

As an implementation choice, it was decided that only the service that issued a Verifiable Credential can revoke it. This is to prevent third parties from revoking Verifiable Credentials that they have not issued. In fact, it is reasonable that only the provider who grants access to the service can eventually revoke it.

As it is possible to notice from the smart contract code, another requirement to be able to add a Verifiable Credential in the smart contract is the fact that it is not already present in the register (line 8), i.e., in the corresponding mapping, there is not a block number indicating which transaction added the credential. If it has not already been added, then it is possible to write it (line 9). When the transaction is successfully added, an event is emitted (line 11), which communicates the issuer of the transaction and the digest of the newly added credential in the register (line 17).

A possible problem is the fact that this smart contract trusts that what is written to the registry is actually a valid digest of a credential in JWT format. In this implementation, there is no kind of access control list that allows only some addresses to write in the smart contract. In fact, once a smart contract is deployed in blockchain, its public methods can be called from any valid address. It is therefore possible that any address can call these methods and write non-consistent information to the register. This problem can be solved with a list of trusted issuers of transactions. In fact, it is possible to consider an issuer as trusted if it also implements the correct cryptographic hash algorithms on the Verifiable Credential before writing it to the register.

- **Verify a Verifiable Credential:**

The verification is the process of evaluation of a Verifiable Credential, in order to determine whether it is authentic and timely valid for the issuer or the presenter. This process includes the following checks:

- the credential conforms to the specification;
- the proof method is satisfied, i.e., the cryptographic mechanism used to prove that the information in a Verifiable Credential was not tampered;
- the credential is not marked as revoked in the smart contract registry.

The Veramo credential library provides the methods for the first two checks, while for the third it is necessary to implement a call to the smart contract registry. The flow for verifying a credential is described in Figure 3.11.

In the implemented solution, in step (1) the data provider calls the Verifiable Credentials micro service, specifying the credential in JWT format to be verified in the request body. Since verifying the presence of a Verifiable Credential on the registry is an operation that does not change the status of the credential, this can be done by any instance of the micro service. In step

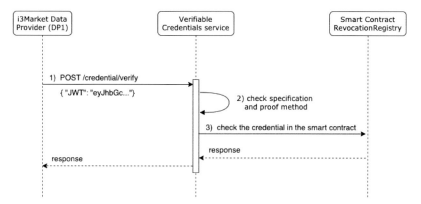

Figure 3.11 Verify Verifiable Credential flow.

(2), the Verifiable Credentials micro service checks that the issuer is valid and that the credential is in a format that complies with the data provider's specifications. If there is no problem with the credential, then it computes the hash of the credential using a SHA-3 cryptographic hash algorithms, which produces a 32-character digest. Then, in step (3), the Verifiable Credential micro service calls the "revoked" method of the smart contract registry, specifying the issuer of the Verifiable Credential and the 32-character digest.

The credential issuer is specified in the JWT and since only the issuer of a credential has the permissions to revoke it, it is sufficient to check that only his address, associated with the credential, is not present in the register. As it is possible to see in the solidity code of the smart contract, detailed in the previous section, this method returns the block number when it was revoked by the "issuer", or 0 if it was not. In this way, it is possible to know if the credential has been revoked or not. This information is then returned as a response to the data provider, who will decide for himself what the next steps will be, for example requesting the issuance of a new valid credential or informing the user that he can no longer request access to that data or service. This API is used in the integration of the OIDC identity provider. In fact, to authenticate a user on the basis of the revealed credentials, a further check on the registry is necessary to ensure that the credential is not revoked.

- **OIDC compatibility:**

The use of Verifiable Credentials allows the distributed and decentralized management of users. In particular, users can use Verifiable Credentials issued as a certificate to obtain a token necessary to access specific services or protected resources within the marketplace. In order to retrieve the Verifiable Credentials and use them in an authorization process, a certified open-source Open ID Connect provider (https://github.com/panva/node-oidc-provider) has been enhanced with wallet API library. In this way, users can be authenticated and authorized based on the Verifiable Credentials they hold as shown in Figure 3.12.

In step (1), the user wants to access a resource or service in the marketplace. The resources and services are made available by data providers, who expect to receive a valid access token and ID token, with the necessary scope to access the resource or service. So, the first thing a data provider website does is to initialize the authentication flow (2). The authentication with authorization code flow + PKCE is done through an OAuth 2.0 SDK (https://github.com/IdentityModel/oidc-client-js), which first generates

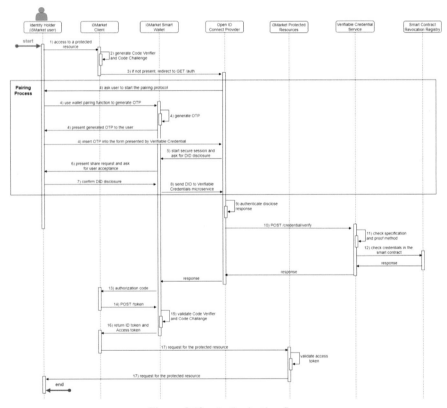

Figure 3.12 Authorization flow.

a code verifier and a code challenge. Specifically, the OAuth 2.0 SDK creates a cryptographically random code verifier and from this generates a code challenge. After that, the authorization code + PKCE flow is initialized with the first call/authorize. The main difference of this Open ID Connect provider compared to traditional ones is that it requires the disclosure of Verifiable Credentials. To specify the credentials to be revealed, the scope field is used.

The Open ID Connect provider has the following static scopes:

- openid: Mandatory for the Open ID Connect standard. It returns the sub-field of the ID token, and its value is the user DID.
- profile: It adds information about the user profile into the ID token.
- vc: It adds the field verifiable claims into the ID token. Useful when the relying party wants to check any information about the verifiable claims asked.

Compared to the standard scope of Open ID Connect, the scopes added are vc and vce. On the other hand, the standard email scope, which returns the user's email, is not present.

There are two different types of scopes:

- vc:vc name: It asks the user for the specific verifiable claim vc name. It is optional; so the user can decide whether to append it or not. If the issuer of the verifiable claim is not trusted, it will be added into untrusted verifiable claims array of the ID token. These arrays are described at the end of this section.
- vce:vc name: It asks the user for the essential verifiable claim with name: vc name. It is mandatory; so if the user does not provide it or the issuer is untrusted, the login and consent process will be cancelled.

After specifying in the scope field which credentials need to be disclosed, the OAuth SDK initializes the authentication process, performing the call to the /authorize endpoint (3).

The Open ID Connect provider performs a selective disclosure request (5), using the Veramo libraries, ask to pair i3-MARKET Smart Wallet, using "pairing" protocol (4).

At this point, a notification will appear on the wallet with the authentication request (Figure 3.13), specifying the credentials that must be revealed (6).

After the disclosure of the required credentials, a callback to the Open ID Connect provider (9) follows. The Open ID Connect checks if all the

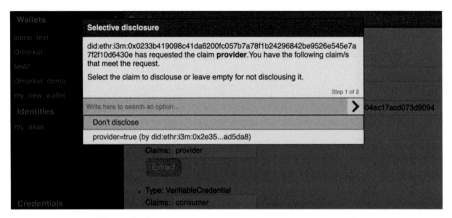

Figure 3.13   Disclosure of the data provider credential.

required Verifiable Credentials are present and if the Verifiable Credential issuer is trusted (10). Subsequently, it remains to check that the credentials are valid and not revoked. In particular, the credentials are sent in JWT format through the verify (11) API, which checks that they have not expired, calculates the hash, and checks if they are present in the Revocation Registry (12). It then returns the array of revoked or invalid credentials as a response. If the response array is empty, then all credentials are valid. If all credentials are valid, the Open ID Connect provider returns the authorization code to the OAuth SDK of the i3-MARKET Data Provider Client (13). The Data Provider SDK performs POST /token (14). The code verifier and code challenge are checked (15) and the ID token and the access token are returned to the Data Provider website (16). Now that the Data Provider website has a valid access token, it can get the resource (17). When the authorization and authentication process finishes, two tokens are returned: access token and ID token. Through the ID token, it is possible to know which of the revealed Verifiable Credentials are verified (trusted) or not (untrusted).

3.4 Diagrams

The following diagrams describe the processes involving the components of the SSI and IAM subsystem.

The diagrams assume that the user created and controlled with his crypto wallet a distributed identity using Ethereum DID management.

3.4.1 Identity authentication

The process in Figure 3.14 describes how a self-sovereign identity is authenticated as managed by a crypto wallet using Ethereum DID management.

The user-centric authentication component create a challengeRequest to retrieve the user's DID and then check the challengeResponse (signed by user's wallet) to verify if the user controls the DID retrieving the corresponding DID document.

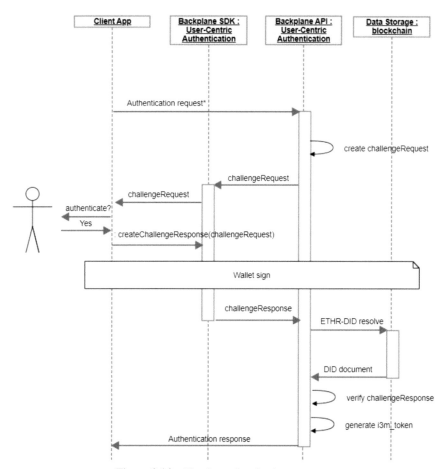

Figure 3.14 Identity authentication process.

3.4.2 User registration

The process illustrated in Figure 3.15 describes how a client application can register a self-sovereign identity as i3-MARKET user issuing a Verifiable Credential attesting his role.

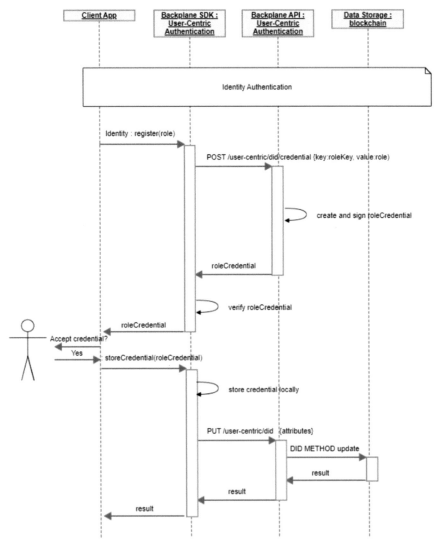

Figure 3.15 User registration process.

After DID authentication and the verification of additional information disclosed by the user, the client app issues a Verifiable Credential for that DID, which attest the role of the user (data consumer, data provider, or both). User's wallet stores the credential locally and update the DID document.

3.4.3 OIDC authorization (authentication code + PKCE)

The following process (Figure 3.16) represents how a client application can be authorized by an i3-MARKET user to access a protected API and obtain information about the user using a standard OpenID Connect Authentication code flow with PKCE.

User-centric authentication component is integrated in a standard OIDC IAM as federated identity provider.

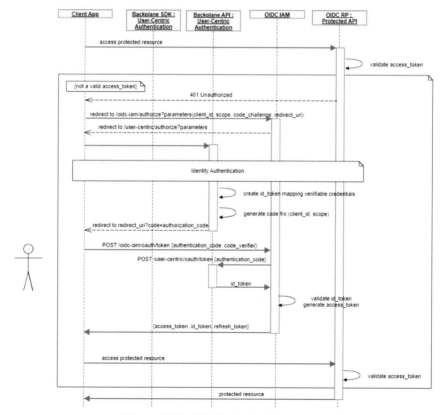

Figure 3.16   OIDC authorization process.

When the client application tries to call a protected resource without a valid access_token, it is redirected to the OIDC IAM authorization endpoint and then to user-centric authentication authorization endpoint showing the login page.

When the user logs in with a wallet, an id_token is created with the DID and the VC associated to the requested scopes and an authentication code is provided to the client to call the token endpoint and receive a valid access_token, a refresh token, and the id_token.

3.5 Interfaces

The interfaces of the final version of Verifiable Credential micro service and OIDC SSI Auth micro service, composing the user-centric authentication component, are presented in Figures 3.17 and 3.18.

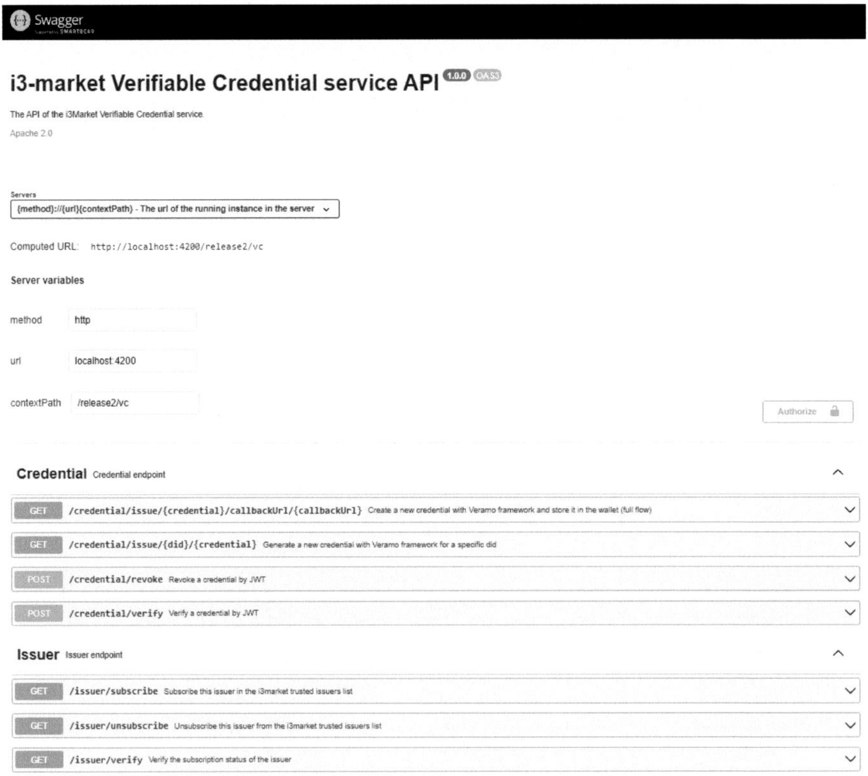

Figure 3.17   Verifiable Credential micro service specification.

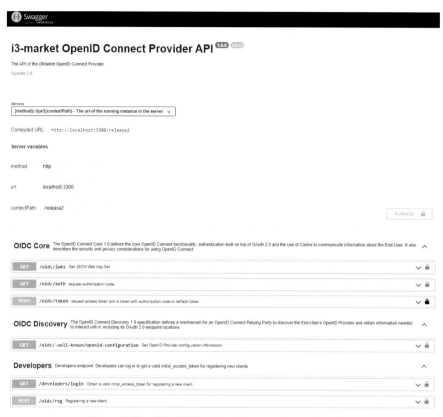

Figure 3.18 OIDC SSI Auth micro service specification.

3.6 Background Technologies

3.6.1 JSON Web Token (JWT)

The JSON Web Token (JWT) is an open standard (RFC 7519) that defines a schema in JSON format for exchanging information between various services. The generated token can be signed (with a secret key that only those who generate the token know) using the HMAC algorithm or using a pair of keys (public/private) using the RSA or ECDSA standards. JWTs are widely used to authenticate requests in Web Services and OAuth 2.0 authentication mechanisms where the client sends an authentication request to the server and the server generates a signed token and returns it to the client who, from that moment on, will use to authenticate subsequent requests. The structure of the token consists of three fundamental parts:

- Header
- Payload
- Signature

The header contains two main information: the type of token (in this case valued to JWT because it is a JSON Web Token) and the type of encryption algorithm used.

```
{
"alg": "HS256",
"typ": "JWT"
}
```

The payload contains the interchange information. It is possible to categorize them into three blocks:

- Registered parameters: They are predefined properties that indicate information about the token (issuer, audience, expiration, issued at, and subject).
- Private parameters: Here, it is possible to enter new fields, such as verifiable claims, having full extensibility, thanks to the JSON structure.
- Public parameters: They refer to parameters defined in the IANA JSON Web Token Registry, and they can be compiled at will by paying attention to the content that is entered to avoid conflicts with the registered and private parameters.

```
{
"iss": "app_name",
"name": "Mario Rossi",
"iat": 1540890704,
exp": 1540918800,
"user": {
"profile": "editor"
}
}
```

The token is generated by encoding the header and payload in base 64 and joining the two results by separating them by a ".", and then the algorithm indicated in the header is applied to the string obtained using a secret key. It is possible to verify and unpack a JWT online using the official website.

Fortunately, it is not necessary to re-implement the encryption logic; there are many libraries to generate JWT depending on the programming language. Security is guaranteed by the fact that the token is signed with a server-side secret key; so if it is corrupted or modified by an external agent, it will not pass validation.

3.6.2 OpenID Connect (OIDC)

OpenID Connect 1.0 (https://openid.net/connect/) is a simple identity layer on top of the OAuth 2.0 protocol. It allows clients to verify the identity of the end-user based on the authentication performed by an authorization server, as well as to obtain basic profile information about the end-user in an interoperable and REST-like manner.

OpenID Connect allows clients of all types, including Web-based, mobile, and JavaScript clients, to request and receive information about authenticated sessions and end-users. The specification suite is extensible, allowing participants to use optional features such as encryption of identity data, discovery of OpenID providers, and session management, when it makes sense for them.

3.6.3 Decentralized identity (DID)

Decentralized identifiers (DIDs) are a new type of identifier that enables verifiable, decentralized digital identity. A DID identifies any subject (e.g., a person, organization, thing, data model, abstract entity, etc.) that the controller of the DID decides that it identifies. In contrast to typical, federated identifiers, DIDs have been designed so that they may be decoupled from centralized registries, identity providers, and certificate authorities. Specifically, while other parties might be used to help enable the discovery of information related to a DID, the design enables the controller of a DID to prove control over it without requiring permission from any other party. DIDs are URIs that associate a DID subject with a DID document allowing trustable interactions associated with that subject.

3.6.4 Self-sovereign identity and blockchain

Today, users' identities and related data are stored in siloes on centralized servers across organizations and are vulnerable to hacking. Repetitive account creation for different applications (e.g., marketplaces), and personal information (often outdated) stored in various services are the disadvantages of that approach.

Self-sovereign identities supported by decentralized systems come as a solution for the following issues:

- Identity and personal data are stored with the user.
- Claims and attestations can be issued and verified between users and trusted parties.
- Users selectively grants access to data.
- Data only needs to be verified a single time.

Blockchain technology, proving decentralization, immutability, and cryptographic security allow the creation of credentials that could be issued and verified without the need of a central certification authority and could be owned by the end-users and directly shared with third parties without involving the credential issuer.

3.6.5 Verifiable Credentials (VC)

As in the physical world, a credential is a set of information that identifies an entity. In particular, the information represents:

- Information related to identifying the subject of the credential (for example, a photo, name, or identification number).
- Information related to the issuing authority (for example, a city government, national agency, or certification body).
- Information related to the type of credential this is (for example, a Dutch passport, an American driving license, or a health insurance card).
- Information related to specific attributes or properties being asserted by the issuing authority about the subject (for example, nationality, the classes of vehicle entitled to drive, or date of birth).
- Evidence related to how the credential was derived.
- Information related to constraints on the credential (for example, expiration date or terms of use).

A Verifiable Credential can represent all of the same information that a physical credential represents (Figure 3.19). The addition of technologies, such as digital signatures, makes Verifiable Credentials more tamper-evident and more trustworthy than their physical counterparts.

Holders of Verifiable Credentials can generate verifiable presentations and then share these verifiable presentations with verifiers to prove they possess Verifiable Credentials with certain characteristics.

Both Verifiable Credentials and verifiable presentations can be transmitted rapidly, making them more convenient than their physical counterparts when trying to establish trust at a distance.

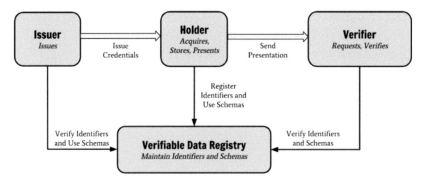

Figure 3.19   Verifiable Credentials model.

Verifiable Credentials are useful in a self-sovereign identity ecosystem because they assert information about the user to whom a credential is issued and can be directly verified by any third-party by involving the issuer. In the context of the project, users are asked to disclose Verifiable Credentials, which attest particular attributes issued by a specific data marketplace. The certified attributes and permissions are used to obtain an OAuth access token that allows the use of Backplane services or access to resources in the marketplace. Verifiable Credentials are therefore used as a proof method in the authorization flow. If a credential is valid, it means that the user is authorized to access a resource or service that requires the holding of that credential. For this reason, a service that generates Verifiable Credentials is necessary. Once a Verifiable Credential is saved by users in their wallets, anyone who receives the Verifiable Credential and has access to the DID of the users can then confirm that the Verifiable Credential has been issued by a trusted server and has not been revoked for some reason.

To implement the solution, we have chosen to use Veramo, a framework that replaces the previous implementation of the uPort library, which is deprecated.

Both components (OIDC SSI Auth and Verifiable Credential micro service) integrate the Veramo framework and take advantage of its features to manage DID and Verifiable Credentials.

4

i3-MARKET Semantic Models

Taking into consideration the nature of the project, we worked on the following:

A. The definition, creation, and collection of semantic data models that allow to share a common description of the data assets (as per the case of shared data offering description model), operations, services, data details, credentials, contracts, pricing, and actors.

B. Development and implementation of semantic engine system and storage for the management of such information, creation of data offering description, management of controlled registries, mapping of information, interfaces among components, links of data and actors, discovery and retrieval of necessary information, compiling of smart contracts, and other operations.

C. Share the semantic models with the community to make use of them and work with people to improve and maintain the models for the present and future.

The use of standardized semantic metadata models and interaction patterns is important to enabling interoperability between nodes, user-friendly services, exchangeability of data assets, representation of actors (marketplaces, providers, consumers, and owners), and data exchange between different instances in the infrastructure ecosystem. A variety of standards already exist for sub-specific topics and domains; the most suitable ones to set up a common information model are selected and integrated into a high-level collection of vocabularies and ontologies. On top of these models, we created i3-MARKET core models to define the missing parts and for the main operational interactions and links among entities. Within i3-MARKET Backplane, the information ecosystem and the infrastructure ecosystem have to be combined to enable a seamless exchange of information and operations in a federated distributed architecture.

From a meta-modelling perspective, the i3-MARKET has raised certain requirements that go beyond the simple main description of datasets, adding information models to define other entities, operations actors, sharing agreements, and data details. While the existing semantic models cover only partially the requirements for the Backplane scopes, we imported, linked, and just in case extended common vocabularies and created the i3-MARKET semantic core model, pricing model, and contractual model for data sharing agreements and service agreements for contracts to compile a collection of semantic information models in O-CASUS models to cover the needs.

i3-MARKET Semantic Models

This section introduces the implementation of i3-MARKET semantic models that comprise the definition, creation, and collection of data models that allow to share a common description of the data assets (as per the case of dynamic data offering descriptions, operations, services, data details, credentials, contracts, pricing, actors, etc.). This section also deals with the definition and implementation of (meta)data management systems and registries catalogues to manage information and meta data descriptions. The main objective of the i3-MARKET Semantic Models is to share the data models with the community to make use of them and work with people on improving and maintaining the models for the present and future.

The i3-MARKET semantic models allow and facilitate the creation of data offering descriptions, management of controlled registries, mapping of information, and distribution of info and details among other components. The models and descriptions provide the links of data assets, metadata, and actors for discovering and retrieving of necessary information, compiling of smart contracts, and other operations.

4.1 i3-MARKET Model Specifications

Specifications for i3-MARKET semantic model solutions comprise the following:

- The definition, creation, and collection of data models that allow to share a common description of the data assets (as per the case of dynamic data offering descriptions, operations, services, data details, credentials, contracts, pricing, actors, etc.).

- The development and implementation of (meta)data management systems and registry catalogues to manage information and metadata descriptions.
- Share the data models with the community to make use of them and work with people on improving and maintaining the models for the present and future.
- The creation of data offering descriptions, management of controlled registries, mapping of information, and interfaces among other components.
- Provide the links of data and actors for discovering and retrieving of necessary information, compiling of smart contracts, and other operations.

We propose the i3-MARKET semantic core model and the semantic models imported and extended that create the collection of O-CASUS models based on the terminologies, definitions, and vocabularies needed to represent the i3-MARKET domain entities and operations. These concepts and their relationships are explained in more detail, including additional sub-concepts.

The O-CASUS semantic models comprise a collection of ontologies and vocabularies to cover the concepts used in the Backplane to define the following:

- i3-MARKET semantic core model
- W3c Data Catalog Vocabulary (DCAT and DCAT-AP)
- W3c Vocabulary of Interlinked Datasets (VoID)
- W3c Verifiable Credentials and DID
- SKOS Simple Knowledge Organization System
- IT Service Ontology
- EU Vocabularies Frequency Named Authority List
- EU Vocabularies File Type Named Authority List
- EU Vocabularies Languages Named Authority List
- EU Vocabularies Continents Named Authority List
- ADMS licence type vocabulary
- Distribution availability vocabulary
- Domain annotations

One of the key aspects when designing a semantic model is the reuse of knowledge. Once a semantic model is created for a domain, it should be (at least to some degree) reusable for other applications in the same domain. To simplify both semantic model development and reuse, a modular design is beneficial. Based on the project specification and the domain environment, the semantic models can be modularized according to their scope, as follows:

- Organization module
- Market module
- Provider module
- Consumer module
- Owner model
- Query module
- Data offering module
- Contractual parameters module
- Data exchange module
- Dataset information module
- Links to pricing module and the other vocabularies and ontologies to cover the various parts of the i3-MARKET O-CASUS sematic information models

A data marketplace is an online transactional location or store that facilitates the buying and selling of data. As many businesses seek to augment or enrich internal datasets with external data, cloud-based data marketplaces are appearing at a growing rate to match data consumers with the right data sellers.

Typical data types for sale in a data marketplace can range from business intelligence and research, demographic, firmographic, and market data to business intelligence and public data. A data marketplace is a more public (and sometimes commercial or monetized) form of data sharing. Data sharing has a long history in academic, research, and public policy circles but in recent years has made enormous inroads into private enterprises, from big business to analyst, consulting, and market intelligence firms. Data consumers include government, analyst, big business, and market intelligence firms. As data volumes continue to explode and machine learning and AI become more important in decision-making, data marketplaces are helping organizations reduce the effort and cost involved in locating required datasets and helping data providers extend their market reach.

However, big data is supported by continuous heterogeneity of underlying data sources (e.g., in IoT spaces), devices and communication technologies, and interoperability in different layers, from communication and seamless integration of platforms to interoperability of data to a global scale.

In a white paper on interoperability [67], it is discussed that many layers of interoperability exist:

- Technical interoperability
- Syntactical interoperability
- Semantic interoperability
- Organizational interoperability
- Dynamic interoperability
- Static interoperability

Discovery, understanding, and collaboration at this level require more than just an ability to interface and to exchange data. Interoperability is "the ability of two or more systems or components to exchange data and use Information" [68], whereas semantic interoperability "means enabling different agents, services, and applications to exchange information, data and knowledge in a meaningful way, on and off the Web" [67][68].

Semantic interoperability is achieved when interacting systems attribute the same meaning to an exchanged piece of data, ensuring consistency of the data across systems regardless of individual data information. This consistency of meaning can be derived from pre-existing standards or agreements on the description and meaning of data or it can be derived in a dynamic way using shared vocabularies either in a schema form or in an ontology-driven approach.

In i3-MARKET, we are aiming at an innovative approach for semantic data, metadata, and modelling activities as represented in Figure 4.1.

To lead the concept of O-CASUS, which is an idea based on the data lifecycle process, we:

- compile vocabularies and taxonomies in relation to marketplaces metadata, operation, and management;
- formalize the state of current marketplaces by using best practices and standards;

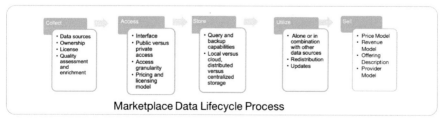

Marketplace Data Lifecycle Process

Figure 4.1 i3-MARKET data model and the data lifecycle process.

- compile an ontology for collecting, accessing, storing, utilizing, and selling data.

4.2 i3-MARKET Semantic Core Models

The Figure 4.2 illustrates the high level of i3-MARKET semantic models that include all the basic conceptual entities and their relationship to all modules.

Details of each module are presented in the following subsections as shown in Figure 4.3. The i3-MARKET Semantic Core Models provide an overview of the i3-MARKET classes of resources that can be members of data offerings and the relationships between them. Except where it does not provide cardinality constraints as they are shown in the Figure 4.4 respectively.

One of the main contributions of the semantic models (vocabularies/ontologies) is the consolidation of the i3-MARKET models and the integration and extensions of other common sematic models to enable the

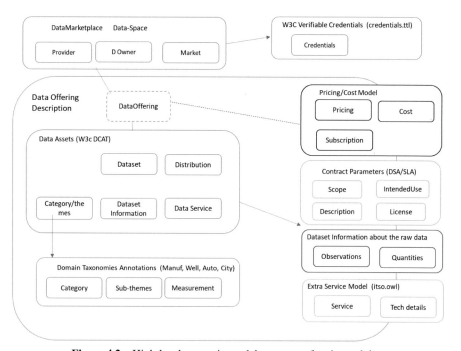

Figure 4.2   High-level semantic model structure of main modules.

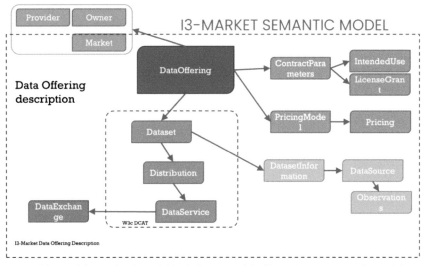

Figure 4.3   Main classes' block diagram of the i3-MARKET semantic model.

mapping of the metadata describing the data assets, contracts, and operations, provided from i3-MARKET stakeholders, to the model/ontology concepts to capture the structural and semantic characteristic of the metadata in relation to the various entities that corresponds to the different data assets and data offerings respectively.

More specifically, the core uses of these models are as follows:

1) Data registration of metadata descriptions, which corresponds to the data harmonization process. In this way, each provided data asset is registered in our registry with concepts from the i3-MARKET data offering model in a semi-automatic way.
2) Metadata linking where any provided data asset metadata will be linked with other relevant sources (or data assets) that exist in the Backplane.
3) Data discovery (for local or federated registries) that involves the development of algorithms and software for supporting the selection of the most appropriate metadata that best match user preferences.
4) Management of information related to smart contract, data access and transfer, pricing models, identity and credential identifications, and notifications.

The i3-MARKET models are used for capturing the structural and semantic metadata characteristics of the various entities involved in the i3-MARKET Backplane domain, whereas the underlying conceptual models

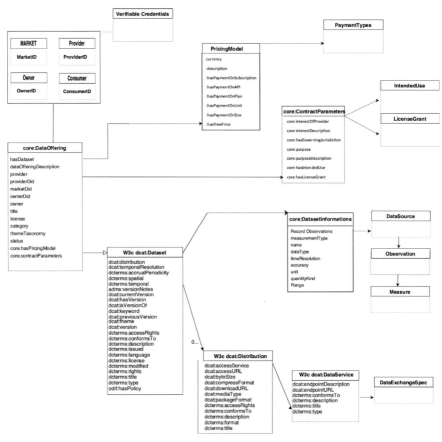

Figure 4.4   Overview of the i3-MARKET semantic model.

facilitate the use of lightweight reasoning during the discovery and operational process, e.g., for contracts and service/agreements, data access/transfer operations, etc.

4.3 Data Marketplace and Data Space Actors

- **Provider module:**

A provider can be a marketplace, data space, or service instance that offers available DataOfferings. A provider is described through the *core:Provider* class. At this stage, each provider has a name and ID (core:providerId) and its organization as shown in Table 4.1. More information about the provider can be added in the future.

Table 4.1 Provider properties.

Property name	Data types	Description
core:providerId	String	Provider ID
providerDescription	String	A description of the provider
providerName	String	Name of the provider
:sourceOrganization	core:Organization	The provider's organization

- **Organization module:**

A provider may also describe its organization. The provider's organization has been an instance of the *schema.org model (Organization Ontology class particularly)*. The connection between the provider and the organization is the sourceOrganization property. Table 4.2 presents some basic properties of the organization class, e.g., at the moment example taken from the *schema.org model and particularly Organization Class.*

Table 4.2 Organization properties.

Property name	Data types	Description
core:organizationId	String	Organization ID
organizationName	String	Name of the organization
:address	String	Physical address of the organization
contactPoint	schema:ContactPoint	A contact point for the organization
organizationDescription	String	A description of the organization

- **Consumer module:**

A consumer can be an entity, application, or service instance that requires access to data resources in order to implement an intended service or function. In the consumer model, we create the *core:Consumer* class that represents the i3-MARKET consumers. Same as the provider, the consumer is also linked to the organization. Table 4.3 presents some basic properties of the core:Consumer.

- **Owner module:**

The actual owner of the data sources provided by marketplace, data space, or service instance that offers available DataOfferings. An owner is

Table 4.3   Consumer properties.

Property name	Data types	Description
core:consumerId	String	Consumer ID
core:dataOfferingQuery	core:DataOfferingQuery	Query to i3-MARKET of consumer
consumerDescription	String	A description of the consumer
consumerName	String	Name of the consumer
:sourceOrganization	schema:Organization	The consumer's organization
core:subscribedTo	core:DataOffering	Data offering IDs the consumer subscribes to

described through the *core:Owner* class. At this stage, each owner has a name (schema:name) and ID (core:ownerID) as shown in Table 4.4. More information about the provider can be added in the future.

Table 4.4   Owner properties

Property name	Data types	Description
core:ownerId	String	Owner ID
ownerDescription	String	A description of the provider
ownerName	String	Name of the provider
:sourceOrganization	schema:Organization	The provider's organization

- **Data market module:**

  Information on the connected data marketplace is given in Table 4.5.

Table 4.5   Market module.

Property name	Data types	Description
core:dataMarketId	String	Data Market ID
dataMarketDescription	String	A description of the data marketplace
dataMarketName	String	Name of the data marketplace
dataMarketNode	String	Info of the data market node

4.4 Data Offering

Data offering description:

The i3-MARKET enables providers to offer or trade access to datasets via the Backplane. A data offering is defined by a "data offering description", which describes via metadata a set of resources offered via the i3-MARKET Backplane. It typically encompasses a set of related information. A data offering description provides a semantic description of the datasets provided to a consumer once the data offering is registered. The description also entails context and meta information about the distribution, including information about the pricing for accessing the resource(s), the license of the information provided, contractual parameters, and service description as URL for data access.

As illustrated in Figure 3.12, the data offering module represents the initial conceptualization, which is built around the DataOffering Class and its metadata. All the core concepts of this module are defined as follows.

A provider registers its offerings on the marketplace by providing an offering description. An offering description is an instance of the data offering class (which can be mapped to the common subclass of schema:Offer). It contains the information about the data assets, data service, categories of data assets, subclass components of catalogues and resources, data services, and categories of the offering (:category). All relevant communication metadata are provided on how the offering can be accessed through the data service and service extension descriptions.

Details of all classes and their properties in the offering module are presented in the following sections.

- **Data Offering description:**

To describe the data assets, contractual parameters, rights, licenses, pricing models, data service, endpoints, format of data, domain annotations, related actors, and other information that describe the datasets, we defined shared "data offering descriptions".

We use W3c Data Catalog Vocabulary (DCAT) – Version 3 vocabulary related to parts such as: dataset, distribution, and DataService used in data offering description (https://w3c.github.io/dxwg/dcat/).

It is recommended to use the description and specifications of DCAT for all the information related to dataset, distribution, and DataService used in data offering description (https://www.w3.org/TR/vocab-dcat-3/).

DCAT enables a publisher to describe datasets and data services in a catalogue using a standard model and vocabulary that facilitates the consumption and aggregation of metadata from multiple catalogues. This can increase the discoverability of datasets and data services. It also makes it possible to have a decentralized approach to publishing data catalogues and makes possible federated search for datasets across catalogues in multiple sites using the same query mechanism and structure.

- **Data Offering class:**

Definition: High-level class in the i3-MARKET core model that introduces the data offering description of dataset resources in Table 4.6.

Table 4.6   Data Offering properties.

Property	Data types	Description
hasDataset	Dataset	Links the data offering in core to a DCAT (-AP) dataset
dataOfferingDescription	String	Contains a free-text account of the *DataOffering*
Provider	Provider	Refers to an entity (organization) responsible for making the data offering available
providerDiD	DID	This is the provider DID, registered in VC and i3-MARKET, which is uniquely identified as a provider in IDM and Wallet
Owner	Owner	Refers to an entity that have source ownership of the data
ownerDiD	DID	This is the owner DID, registered in VC and i3-MARKET, which is uniquely identified as an owner in IDM and Wallet
Marketid	Market	This is the market name ID, which is uniquely identified as a marketplace
marketDiD	DID	This is the market DID, registered in VC and i3-MARKET, which is uniquely identified as a marketplace in IDM and Wallet
dataOfferingTitle	String	Contains a name given to the catalogue
License	LicenseDocument	This property refers to the license under which the catalogue can be used or reused
Category	ConceptScheme	Refers to a knowledge organization system used to classify the data offering categories for datasets The high-level category terms and the URI used are defined in the scheme file DataOfferingCategory.ttl
themeTaxonomy	skos:ConceptScheme	This property refers to a knowledge organization system used to classify the DataOffering's datasets

Table 4.6 *Continued.*

Property	Data types	Description
Active	String	Flag to set if the DataOffering is activated/available by the provider to be checked/searched by, e.g., the consumer
core:ownerConsentForm		Hashtag string to report the information about the explicit user consent form documentations
core:inSharedNetwork		Boolean to define if the DataOffering is shared by the marketplace to be visible and consumable by all actors in the i3-MARKET network
core:status		To define the data offering status
core:dataOffering ExpirationTime		Expiration time of DataOffering in case
core:lastModified		Most recent date on which the data offering was changed, updated, or modified
dcat:previousVersion		The previous version of a resource in a lineage (PAV)
Version		To define the "version" of the registered data offering
Core:datasetInformation		For the module that describes the information related to the details of the raw data (with info on origin of data, measurements, data types, devices, units, etc.)
core:hasPricingModel	pricingmodel: PricingModel	The pricing model for the data offering
core:contractParameters	core: ContractParameters	Some specific contract parameters related to data offering

- **(DCAT) Dataset class:**

Definition: A collection of data published or curated by a single agent and available for access or download in one or more representations, as shown in Table 4.7.

- **DatasetInformation class:**

Definition: Extended specific annotations to add extra information related to a dataset. This information is used to give providers the possibility to describe with more granularity the source and types of data in datasets

Table 4.7   DCAT dataset main properties.

Property	Data types	Description
Description	String	Contains a free-text account of the dataset
Title	String	Contains a name given to the dataset
Keyword	String	Contains a keyword or tag describing the dataset
core:datasetInformation	core:DatasetInformation	Some specific information annotations of dataset metadata information types, which represent attributes of observations, measurements, fields, etc. in the dataset
core:datasetRecord	core:DatasetRecord	In case data records types that represent attributes of fields, in the dataset
datasetDistribution	Distribution	Links the dataset to an available distribution
geographicalCoverage	Location	Refers to a geographical area covered by the dataset
temporalCoverage	PeriodOfTime	Refers to a temporal period that the dataset covers
Category	Concept	Refers to a category of the dataset. A dataset may be associated with multiple categories
accessRights	RightsStatement	Refers to information that indicates whether the dataset is open data, has access restrictions, or is not public
Frequency	Frequency	Refers to the frequency at which the dataset is updated
Documentation	Documentation	Refers to a page or document about this dataset
hasVersion	Dataset	Refers to a related dataset that is a version, edition, or adaptation of the described dataset
Creator	Agent	Refers to the entity primarily responsible for producing the dataset
dcat:theme	skos:Concept	This property refers to a category of the dataset. A dataset may be associated with multiple themes

and annotations related to specific domains (see Table 4.8). Also consult Appendix B for an extended version of information details to be used to describe the raw original data for consumers' understanding).

Table 4.8 Main properties of the DatasetInformation class.

Property	Data types	Description
core:measurementType	String	The data types that represent attributes of observations, measurements in the dataset
core:measurementChannelType	String	The data measurement channel types in the dataset
core:sensorID	String	Sensor ID
core:deviceID	String	Device ID
core:cppType	String	Cyber−physical systems cpp type
core:sensorType	String	Sensor type

- **(DCAT) Distribution class:**

   Definition: A specific representation of a dataset in Table 4.9.

Table 4.9 DCAT distribution main properties.

Property	Data types	Description
Description	String	Contains a free-text account of the distribution
accessURL	Resource	Contains a URL that gives access to a distribution of the dataset
Availability	Concept	Indicates how long it is planned to keep the distribution of the dataset available
Format	MediaTypeOrExtent	Refers to the file format of the distribution
downloadType	String	Download type (it means frequency as "Stream" or "Bulk" dataset can be downloaded)
License	LicenseDocument	Refers to the licence under which the distribution is made available
accessService	DataService	Refers to a data service that gives access to the distribution of the dataset
byteSize	Double	Size of a distribution in bytes
Documentation	Documentation	Refers to a page or document about this distribution
downloadURL	Resource	URL that is a direct link to a downloadable file in a given format
releaseDate	DateTime	Contains the date of formal issuance (e.g., publication) of the distribution

- **(DCAT) DataService class:**

Definition: A collection of operations that provides access to one or more datasets or data processing functions is shown in Table 4.10.

Table 4.10   DataService properties.

Property	Data types	Description
Description	String	Contains a free-text account of the data service
endpointURL	Resource	The root location or primary endpoint of the service (an IRI)
Title	String	Contains a name given to the data service
servesDataset	Dataset	Refers to a collection of data that this data service can distribute
License	LicenseDocument	Refers to the licence under which the data service is made available
accessRights	RightsStatement	Includes information regarding access or restrictions based on privacy, security, or other policies
serviceID	String	Service ID
serviceSpecs	ServiceSpecs	Service specification reference to ITSO extra service model specifications

- **ContractParameters class:**

Definition: A collection of parameters that provides information about the use and scope of the DataOffering/dataset in Table 4.11.

Table 4.11   DataService properties.

Property name	Data types	Description
core:interestOfProvider	Literal	This property is used to identify the interest of the data owner. The following possibilities exist: Free sharing quotation; selling of data (e.g., just earning money by selling the data, no specific feedback on these data by a data consumer expected)
core:interestDescription	Literal	Data provider can specify which sort of quotation he wants exactly, e.g., quotation for maintenance service or quotation for optimization of production
core:hasGoverningJurisdiction	Literal	Jurisdiction

Table 4.11 *Continued.*

Property name	Data types	Description
core:purpose	Literal	Purpose for the use of the dataset
core:purposeDescription	Literal	Description of the purpose for the use of the dataset
core:hasIntendedUse	core:IntendedUse	To intended use class/properties
core:hasLicenseGrant	core:LicenseGrant	To license grant class/properties

- **LicenseGrant class:**

Definition: Definition of the type of license is associated with the data asset in Table 4.12.

Table 4.12 LicenseGrant properties.

Property name	Data/object types	Description
core:paidUp	Boolean	If licence grant to paidUp
core:transferable	Boolean	Transferable (true or false)
core:exclusiveness	Boolean	License of exclusiveness (true or false)
core:revocable	Boolean	License revocable (true or false)
core:processing	Boolean	If licence grant data to be processed
core:modifying	Boolean	If licence grant data to be modified
core:analyzing	Boolean	If licence grant data to be analysed
core:storingData	Boolean	If licence grant to store data
core:storingCopy	Boolean	If licence grant to store copy of data
core:reproducing	Boolean	If licence grant to reproduce data
core:distributing	Boolean	If licence grant to distribute data
core:loaning	Boolean	If licence grant to loan data
core:selling	Boolean	If licence grant to sell data
core:renting	Boolean	If licence grant to rent data
core:furtherLicensing	Boolean	If licence grant for further licensing
core:leasing	Boolean	If licence grant to lease data

- **IntendedUse class:**

Definition: What the data provider allows the consumer to be the intended use of the data assets in Table 4.13.

Table 4.13   IntededUse properties.

Property name	Data/object types	Description
core:processData	Boolean	Process data (true or false)
core:shareDataWithThirdParty	Boolean	Share data with third party (true or false)
core:editData	Boolean	Edit data (true or false)

- **DataExchangeSpec class:**

Definition: Information inside the accessService block for data exchange specifications that serve the distributions of the datasets, used also by the data access and transfer system in Table 4.14.

Table 4.14   DataExchange properties.

Property name	Data/object types	Description
core:encAlg	Boolean	Encryption algorithm used to encrypt blocks. Either AES-128-GCM ('A128GCM') or AES-256-GCM ('A256GCM')
core:signingAlg	Boolean	Signing algorithm used to sign the proofs. Like ECDSA secp256r1 with key lengths: either "ES256", "ES384", or "ES512"
core:hashAlg	Boolean	Hash algorithm used to compute digest/commitments. It is SHA2 with different output lengths: either "SHA-256", "SHA-384", or "SHA-512"
core:ledgerContractAddress		The ledger smart contract address (hexadecimal) on the DLT
core:ledgerSignerAddress		The orig (data provider) address in the DLT (hexadecimal)
core:pooToPorDelay		Maximum acceptable delay between the issuance of the proof of origin (PoO) by the orig and the reception of the proof of reception (PoR) by the orig
core:pooToPopDelay		Maximum acceptable delay between the issuance of the proof of origin (PoO) by the orig and the reception of the proof of publication (PoP) by the dest
core:pooToSecretDelay		If the dest (data consumer) does not receive the PoP, it could still get the decryption secret from the DLT. This defines the maximum acceptable delay between the issuance of the proof of origin (PoO) by the orig and the publication (block time) of the secret on the blockchain

For a more complete list of classes and attributes that are used for the data offering description and details on their definitions, please see Tables 4.15 and 4.16.

Table 4.15 Preliminary example for a metadata description.

```
##
PREFIX rdfs: <http://www.w3.org/1999/02/22-rdf-syntax-ns#>
PREFIX rdf: <http://www.w3.org/2000/01/rdf-schema#>
PREFIX xsd:  <http://www.w3.org/2001/XMLSchema#>
PREFIX dct:  <http://purl.org/dc/terms/>
PREFIX dcat: <http://www.w3.org/ns/dcat#>
PREFIX pricingmodel: <http://i3-MARKET.eu/Backplane/pricingmode/>
PREFIX core: <http://i3-MARKET.eu/Backplane/core/>
PREFIX : <http://i3-MARKET.org/resource/>

:Mindsphere
  a core:Provider ;
  rdfs:label "Mindsphere"@en ;
  core:dataOffering ex:DataOffering-1  .
```

Table 4.16 Preliminary example for a DataOffering description.

```
Preliminary example for a DataOffering description
{
"@context": {
    "core": "http://i3-MARKET.eu/Backplane/core/"
    "dcat": "https://www.w3.org/ns/dcat.jsonld"
    "pricingmodel": "http://i3-MARKET.eu/Backplane/pricingmodel"
},
"id":       "#####-#######-#######-###"        OR        "http://i3-
MARKET.org/resource/#####-#######-#######-###"
"type": "http://i3-MARKET.eu/Backplane/core/DataOffering"

"provider": "#####-#######-#######-###"
"marketId": "#####-#######-#######-###",
"owner": "#####-#######-#######-###",
"dataOfferingTitle": "_field",
"dataOfferingDescription": "string",
"category": "Other",
"status": "e.g. Activated, InActivated, ToBeDeleted, Deleted",
"dataOfferingExpirationTime": "NA",
"contractParameters":
    {
    "id": "http://i3-MARKET.org/resource/#####-#######-#######-###"
    "type": "http://i3-MARKET.eu/Backplane/core/ContractParameters"
```

Table 4.16   *Continued.*

```
"contractParametersId": "string",
"interestOfProvider": "NA",
"interestDescription": "NA",
"hasGoverningJurisdiction": "NA",
"purpose": "NA",
"purposeDescription": "NA",
"hasIntendedUse":
    {
      "id": "http://i3-MARKET.org/resource/#####-#######-#######-###"
      "type": "http://i3-MARKET.eu/Backplane/core/IntendedUse"

      "intendedUseId": "string",
      "processData": "true OR false",
      "shareDataWithThirdParty": "true OR false",
      "editData": "true OR false"
    }       ,
"hasLicenseGrant":
    {
      "id": "http://i3-MARKET.org/resource/#####-#######-#######-###"
      "type": "http://i3-MARKET.eu/Backplane/core/LicenseGrant"

      "licenseGrantId": "string",
      "copyData": "true OR false",
      "transferable": "true OR false",
      "exclusiveness": "true OR false",
      "revocable": "true OR false"
    }
  }   ,

"hasDataset":
  {
    "id": "http://i3-MARKET.org/resource/#####-#######-#######-###"
    "type": "http://www.w3.org/ns/dcat#Dataset"

    "datasetId": "string",
    "title": "_field",
    "keyword": "_field",
    "dataset": "_field",
    "description": "_field",
    "issued": "date-time",
    "modified": "date-time",
    "temporal": "_field",
    "language": "_field",
    "spatial": "_field",
    "accrualPeriodicity": "_field",
    "temporalResolution": "_field",
    "distribution": [
        {
          "id": "http://i3-MARKET.org/resource/#####-#######-#######-###"
          "type": "http://www.w3.org/ns/dcat#Distribution"

          "distributionId": "string",
          "title": "_field",
          "description": "_field",
          "license": "_field",
          "accessRights": "_field",
```

Table 4.16 *Continued.*

```
            "downloadType": "_field",
            "conformsTo": "_field",
            "mediaType": "_field",
            "packageFormat": "_field",
            "accessService":
                {
                    "id":    "http://i3-MARKET.org/resource/#####-#######-#######-
      ###"
                    "type": "http://www.w3.org/ns/dcat#DataService"

                    "dataserviceId": "string",
                    "conformsTo": "_field",
                    "endpointDescription": "_field",
                    "endpointURL": "_field",
                    "servesDataset": "_field",
                    "serviceSpecs": "_field"
                }
        }
    ],
    "datasetInformation": [
        {
            "id": "http://i3-MARKET.org/resource/#####-#######-#######-###"
            "type": "http://i3-MARKET.eu/Backplane/core/DatasetInformation"

            "datasetInformationId": "string",
            "measurementType": "_field",
            "measurementChannelType": "_field",
            "sensorId": "_field",
            "deviceId": "_field",
            "cppType": "_field",
            "sensorType": "_field"
        }
    ],
    "theme": [
        "_field"
        "_field"
        "_field"
    ]
    }
}
```

4.4.1 Controlled vocabularies suggested to be used for particular annotations

In Table 4.17, a number of properties are listed with controlled vocabularies that should be used for the listed properties. The declaration of the following controlled vocabularies as high recommendation (in DCAT_AP specifications are listed as mandatory) ensures a minimum level of interoperability.

Table 4.17   Preliminary example for annotations.

Property URI	Used for class	Vocabulary name	Vocabulary URI	Usage note
dcat:mediaType	Distribution	IANA Media Types [5]	http://www.iana.org /assignments/media -types/media-types .xhtml	
dcat:theme	Dataset	Dataset Theme Vocabulary	http://publications .europa.eu/resourc e/authority/data-the me	The values to be used for this property are the URIs of the concepts in the vocabulary
dcat:themeTaxonomy	Catalogue	Dataset Theme Vocabulary	http://publications.e uropa.eu/resource/ dataset/data-theme	The value to be used for this property is the URI of the vocabulary itself, i.e., the concept scheme, not the URIs of the concepts in the vocabulary
dct:accrualPeriodicity	Dataset	EU Vocabularies Frequency Named Authority List [70]	http://publications.e uropa.eu/resource/ authority/frequency	
dct:format	Distribution	EU Vocabularies File Type Named Authority List [71]	http://publications.e uropa.eu/resource/ authority/file-type	
dct:language	Catalogue, dataset, catalogue record, distribution	EU Vocabularies Languages Named Authority List [72]	http://publications.e uropa.eu/resource/ authority/language	
dct:publisher	Catalogue, dataset	EU Vocabularies Corporate Bodies Named Authority List [73]	http://publications .europa.eu/resourc e/authority/corporat e-body	The corporate bodies NAL must be used for European institutions and a small set of international organizations. In case of other types of organizations, national, regional, or local vocabularies should be used
dct:spatial	Catalogue, dataset	EU Vocabularies Continents Named Authority List [74], EU Vocabularies Countries Named Authority List [75], EU Vocabularies Places Named Authority List [76], Geonames	http://publications.e uropa.eu/resource/ authority/continen t/,~http://publicatio ns.europa.eu/resou rce/authority/coun try,~http://publicat ions.europa.eu/res ource/authority/pl ace/, http://sws.geonam es.org/	The EU Vocabularies Name Authority Lists must be used for continents, countries, and places that are in those lists; if a particular location is not in one of the mentioned Named Authority Lists, Geonames, URIs must be used
adms:status	Distribution	ADMS status vocabulary	http://purl.org/adm s/status/	The list of terms in the ADMS status vocabulary is included in the ADMS specification [77]
dct:type	Agent	ADMS publisher type vocabulary	http://purl.org/adm s/publishertype/	The list of terms in the ADMS publisher type vocabulary is included in the ADMS specification

Table 4.17   *Continued.*

Property URI	Used for class	Vocabulary name	Vocabulary URI	Usage note
dct:type	Licence document	ADMS licence type vocabulary	http://purl.org/adm s/licencetype/	The list of terms in the ADMS licence type vocabulary is included in the ADMS specification
dcatap:availability	Distribution	Distribution availability vocabulary	http://data.europa. eu/r5r/availability/	The list of terms for the availability levels of a dataset distribution is included in the DCAT-AP specification

4.4.2 Pricing model

Here we present the general representation of a pricing model to describe the pricing information attached to the data assets related to legacy information of pricing specification in the marketplaces.

Pricing models associated with the DataOffering class is shown in Table 4.18.

- **Base class pricingmodel:PricingModel:**

Table 4.18   PricingModel basic properties.

Property name	Data types	Description
currency	xyz	Currency type
:description	String	A description
:name	String	Name
:hasPaymentOnSubscription	pricingmodel: PaymentOnSubscription	PaymentOnSubscription
:hasPaymentOnAPI	pricingmodel:PaymentOnAPI	PaymentOnAPI
:hasPaymentOnPlan	pricingmodel:PaymentOnPlan	PaymentOnPlan
:hasPaymentOnUnit	pricingmodel:PaymentOnUnit	PaymentOnUnit
:hasPaymentOnSize	pricingmodel:PaymentOnSize	PaymentOnSize
:hasFreePrice	pricingmodel:FreePrice	FreePrice

For payment categories from marketplace terms, we can have like:
pricingmodel:PaymentOnPlan, pricingmodel:PaymentOnAPI,
pricingmodel:PaymentOnUnit,
pricingmodel:PaymentOnSize,
pricingmodel:PaymentOnSubscriptiOn, and pricingmodel:FreePrice.

- **PaymentOnPlan class:**

The payment type class pricingmodel:PaymentOnPlan is shown in Table 4.19.

Table 4.19   PaymentOnPlan basic properties.

Property name	Data types	Description
:hasPlanPrice	String	Plan price
:description	String	A description
:name	String	Name
: :planDuration	String	Plan duration

- **PaymentOnAPI class:**

The payment type class pricingmodel:PaymentOnAPI is shown in Table 4.20.

Table 4.20   PaymentOnAPI basic properties.

Property name	Data types	Description
:hasAPIPrice	String	Basic price
:description	String	A description
:name	String	Name
pricingmodel:numberObject		Number of objects moved via API

- **PaymentOnUnit class:**

The payment type class pricingmodel:PaymentOnUnit is shown in Table 4.21.

Table 4.21   PaymentOnUnit basic properties.

Property name	Data types	Description
:hasUnitPrice	String	Basic price
:description	String	A description
:name	String	Name
pricingmodel:dataUnit		Data unit type
:unitID	String	:unit ID

- **PaymentOnSize class:**

The payment type class pricingmodel:PaymentOnSize is shown in Table 4.22.

Table 4.22 PaymentOnSize basic properties.

Property name	Data types	Description
:hasSizePrice	String	Basic price
:description	String	A description
:name	String	Name
:dataSize		Size of data

Table 4.23 PaymentOnSubscription basic properties.

Property name	Data types	Description
:hasSubscriptionPrice	String	Subscription price
:description	String	A description
:name	String	Name
:timeDuration	Time	Subscription d duration
:fromValue	Date time	Subscription validity starting point
:toValue	Date time	Subscription validity ending point
:repeat	pricingmodel:RepeatBy	In case the subscription is repeatable

- **PaymentOnSubscriptiOn class:**

The payment type class pricingmodel:PaymentOnSubscriptiOn is shown in Table 4.23.

- **FreePrice class:**

The payment type class pricingmodel:FreePrice is shown in Table 4.24.

Table 4.24 FreePrice basic properties.

Property name	Data types	Description
:hasFreePrice	String	Free option

Suggested data pricing-value model:

The pricing information is useful to compile the necessary details for smart contracts and other auditable information (Figure 4.5), plus we use parameters for helping users evaluate the possible best suggested prices for their assets that have to be traded/shared (Figure 4.6).

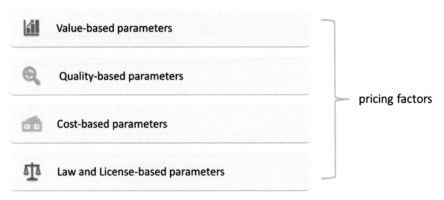

Figure 4.5   Data pricing-value model.

Parameters used in our price recommendation calculator tool are shown in Figure 4.6.

Parameters	Self-Score	Weight
Cost of collecting, storing or/and analysis	X €	Y
Estimated data value for the consumer	X €	Y
Data Completeness	X	Y
Data Accuracy and Validity	X	Y
Unique entries/values	X	Y
Data Rarity/Scarceness	X	Y
Type of license	X	Y
Credibility of the seller (*IF available*)	X	Y
Price of the dataset	**X €**	
Recommended Price of the dataset (including credibility of seller *if available*)	**X €**	

Figure 4.6   Data pricing-value model parameters.

4.4.3 Domain categorization/taxonomies for domain-specific annotations of datasets

Property: core:category and dcat:theme:

The dcat:theme is used to give annotation and information about the domain categorization of the datasets. In i3-MARKET, we use the themes as sub-categories to give more granularity in defining the domain annotations. In DCAT 1, the domain of *dcat:theme* was *dcat:Dataset*, which limited the use of this property in other contexts. The domain has been relaxed in later revisions.

We also added a upper level property for a data offering to annotate directly the high-level type of category the data offering belongs to as core:category.

Semantic attribute:	dcat:theme
Definition:	A category of the resource. A resource can have multiple themes
Sub-property of:	dct:subject
Range:	skos:Concept
Usage note:	The set of skos:Concepts used to categorize the resources are organized in a skos:ConceptScheme describing all the categories and their relations in the catalogue

Class: ConceptScheme:

Semantic class:	skos:ConceptScheme
Definition:	A knowledge organization system (KOS) used to represent themes/categories of datasets in the catalogue

Class: Concept:

Semantic Class:	textbfskos:Concept
Definition:	A category or a theme used to describe datasets in the catalogue
Usage note:	It is recommended to use either skos:inScheme or skos:topConceptOf on every skos:Concept used to classify datasets to link it to the concept scheme it belongs to. This concept scheme is typically associated with the catalogue using dcat:themeTaxonomy

We are using skos:ConceptScheme via skos:Concept to create taxonomies to annotate high-level types of annotations for domain themes/categories classifications.

Example of category terms as in i3-MARKET DataCategory.ttl schema.

```
Data Categories

<http://i3.market.eu/auth/dataCatagory/Manufacturing>
            skos:prefLabel "Manufacturing"@en.
<http://i3.market.eu/auth/dataCatagory/Automotive>
            skos:prefLabel "Automotive"@en.
<http://i3.market.eu/auth/dataCatagory/Wellbeing>
            skos:prefLabel "Wellbeing"@en.
<http://i3.market.eu/auth/dataCatagory/Agriculture>
            skos:prefLabel "Agriculture,  fisheries,  forestry  and
food"@en.
<http://i3.market.eu/auth/dataCatagory/Culture>
            skos:prefLabel "Culture and sport"@en.
<http://i3.market.eu/auth/dataCatagory/Economy>
            skos:prefLabel "Economy and finance"@en.
<http://i3.market.eu/auth/dataCatagory/Education>
            skos:prefLabel "Education"@en.
<http://i3.market.eu/auth/dataCatagory/Energy>
            skos:prefLabel "Energy"@en.
<http://i3.market.eu/auth/dataCatagory/Environment>
            skos:prefLabel "Environment"@en.
<http://i3.market.eu/auth/dataCatagory/Government>
            skos:prefLabel "Government and public sector"@en.
<http://i3.market.eu/auth/dataCatagory/Health>
            skos:prefLabel "Health"@en.
<http://i3.market.eu/auth/dataCatagory/International>
            skos:prefLabel "International issues"@en.
<http://i3.market.eu/auth/dataCatagory/Justice>
            skos:prefLabel "Justice, legal system and public safety"@en.
<http://i3.market.eu/auth/dataCatagory/Regions>
            skos:prefLabel "Regions and cities"@en.
<http://i3.market.eu/auth/dataCatagory/society>
            skos:prefLabel  "Population and society"@en.
<http://i3.market.eu/auth/dataCatagory/Science>
            skos:prefLabel "Science and technology"@en.
<http://i3.market.eu/auth/dataCatagory/Transport>
            skos:prefLabel "Transport"@en.
```

4.4.4 W3C Verifiable Credentials data model

For representing the Verifiable Credentials, the Backplane follows the W3c Verifiable Credentials Data Model 1.0.

Credentials are a part of our daily lives; driver's licenses are used to assert that we are capable of operating a motor vehicle, university degrees can be used to assert our level of education, and government-issued passports enable us to travel between countries. These credentials provide benefits to us when used in the physical world, but their use on the Web continues to be elusive.

Currently, it is difficult to express education qualifications, healthcare data, financial account details, and other sorts of third-party verified machine-readable personal information on the Web. The difficulty of expressing digital credentials on the Web makes it challenging to receive the same benefits through the Web that physical credentials provide us in the physical world.

This specification provides a standard way to express credentials on the Web in a way that is cryptographically secure, privacy respecting, and machine verifiable.

Also, in i3-MARKET, the SSI& IAM subsystems use DIDs that follow the W3c decentralized identifiers (DIDs) v1.0 specifications.

Decentralized identifiers (DIDs) are a new type of identifier that enables verifiable, decentralized digital identity. A DID refers to any subject (e.g., a person, organization, thing, data model, abstract entity, etc.) as determined by the controller of the DID. In contrast to typical, federated identifiers, DIDs have been designed so that they may be decoupled from centralized registries, identity providers, and certificate authorities. Specifically, while other parties might be used to help enable the discovery of information related to a DID, the design enables the controller of a DID to prove control over it without requiring permission from any other party. DIDs are URIs that associate a DID subject with a DID document allowing trustable interactions associated with that subject.

4.4.5 Smart contracts for data sharing agreements

How to create a fair and trusted ecosystem around the sharing/trading of data?

✓ Contracts set the basis of the data/sharing/trading.
✓ Contractual agreements are critical to ensure the fair and trustworthy sharing/trading of data.
✓ Smart contracts offer new affordances and opportunities to enhance trust in trading and sharing of data.

Most common clauses for DSAs:

○ General legal provisions:
  ■ Parties and roles.
  ■ Preamble, scope of the agreement, and definitions.

- ■ Description/specification of the subject matter > description/scoping of the data and data trading and the agreement.
- ■ Duties/obligations and rights > terms & conditions for data trading (room for flexibility).
- ■ Intended use.
- ■ Warranties > room for flexibility.
- ■ Liability and dispute resolution mechanisms > room for flexibility.
- ■ Duration and termination > performance, breach, modification, conditions for premature termination, and consequences.

○ Specific license types:

- ■ License grants.
- ■ Intended use.

4.5 Online i3-MARKET Semantic Model Repository and Community Management

The results are shared not only with project partners but also with stakeholders and community in open-source repositories. As part of open-source assets, the data models, documentations, and files used in the i3-MARKET project are made available, such as the following.

- • The i3-MARKET data pack is the set of files, schemas, and metadata model diagrams that represent the way the i3-MARKET semantics are organized and structured; it also contains the metadata in two different formats, e.g., ttl and Jason-ld. owl.
- • The i3-MARKET semantic model info is the documentation that describes in detail all the taxonomies and vocabularies from needed domains used in i3-MARKET and that describes and represents all the relationships between them to build the graph representation of the i3-MARKET semantic model.
- • The support repo is the mechanism for how the data model is maintained following the interoperability requirements in i3-MARKET. If you want to contribute or have any suggestion for improving the semantic models, visit the open-source repositories and contact the authors and members.
- • The model files are shared in i3-MARKET GitHub/Gitlab repositories with release versions where each section contains the online machine-readable files in OWL and other format for online accessibility. The files

are maintained and updated regularly to keep the latest version of the models files up to date.

The code as well the models and vocabularies are available open-source via the establishment of the i3-MARKET spaces on Gitlab available at: https://gitlab.com/i3-MARKET-V3-public-repository/, and GitHub available at: https://github.com/i3-MARKET-V3-public-repository/.

The i3-MARKET semantic models governance process, which is defined as the support and evaluation process to include semantic improvements, is as follows:

- **Request for changes or updates:** Identify any changes previous to a major release, which should be considered private and usually is on testing and pre-consensus/staging.
- **The evaluation of any type of update request:** A review from editors and community approves participation and updates. In particular terms, vocabularies, ontologies or initiate a model in i3-MARKET OSS.
- **The communication of the results from technical experts:** A tagging version using alpha, beta, and gamma versions and then tagged as major is used here.
- **Evaluation of contributions for new commits:** Technical experts, PM, TM, TPMs, WPLs, and TaskLs, Assess and Evaluate the Contribution Includes documentation at the initiated project in i3-MARKET OSS.
- **Reports and changes report:** The technical board issues a short report, explaining the rational on the rejection in exceptional cases. This step can include rejection/cancellation of project participation.

It is possible to find a more complete definition of the attributes used in the data offerings description schema template as used in the Semantic Engine API in Appendix A.

4.6 Data Offerings Description–Schema Definitions in the API Template

When creating resources as per the data offering description, you fill the attribute fields to describe the traded/shared assets and datasets in the templates that are registered in the registry catalogue and allow the collection of information that are used by the engine and other components to retrieve details for search and retrieval of data for information systems and operations.

You can find the main semantic data model files for i3-MARKET in open-source GitHub and Gitlab repository project, e.g., at https://github.com/i3-MARKET-V3-Public-Repository/SemanticsDataModels.

Definitions for semantic description of data offerings in relation to the API template:

DataOffering:

{

"marketId":

Semantic attribute:	core:marketId
Definition:	This is the market name ID, which is uniquely identified a marketplace
Range:	Marketplace identifier: xsd:string
Usage note:	n/a
See also:	n/a

"provider":

Semantic attribute:	core:provider
Definition:	Provider of the DataOffering
Range:	Provider identifier: xsd:string
Usage note:	Should be the identifier of the provider in the i3-MARKET system verification that should be done with registered providers. All other providers shall be rejected.Return an error message in case an unregistered provider is specified.
See also:	Maybe connected with the IDs in identity manager. As the actual registration is by the Marketplaces/DataSpaces, they have the knowledge and responsibility to have the name/identity of the providers (that have knowledge of the owners) whom they would know are the providers

"owner":

Semantic attribute:	core:owner
Definition:	Owner of the DataOffering
Range:	Owner identifier: xsd:string
Usage note:	Should be the identifier of the owner in the i3-MARKET system. Owners are not registered in i3-MARKET. Optional parameter. Not to be verified.
See also:	n/a

"marketDid": (could be automatically filled by, e.g., WEB-RI in the creation moment of the data offering)

Semantic attribute:	core:marketDid
Definition:	This is the market DID, registered in VC and i3-MARKET, which is uniquely identified a marketplace
Range:	Marketplace identifier: DID
Usage note:	This ID is generated at the marketplace level, and inserting into an offering automatically by the marketplace itself rather than by a user.
See also:	

"providerDid": (could be automatically filled by, e.g., WEB-RI in the creation moment of the data offering)

Semantic attribute:	core:providerDid
Definition:	Provider of the DataOffering DID, registered in VC and i3-MARKET, which is uniquely identified
Range:	Provider identifier: DID
Usage note:	Should be the identifier of the provider in the i3-MARKET system. Verification should be done with registered providers. All other providers shall be rejected.Return an error message in case an unregistered provider is specified.
See also:	linked to VC

"ownerDid": (could be automatically filled by, e.g., WEB-RI in the creation moment of the data offering)

Semantic attribute:	core:ownerDid
Definition:	Owner of the DataOffering DID, registered in VC and i3-MARKET, which is uniquely identified
Range:	Owner identifier: DID
Usage note:	Should be the identifier of the owner in the i3-MARKET system. Owners are not registered in i3-MARKET. Optional parameter. Not to be verified.
See also:	Maybe connected with the IDs in identity manager

"ownerConsentForm": (should be implemented allowing the indication for user consent form hash details)

Semantic attribute:	core:ownerConsentForm
Definition:	Hashtag string to report the information about the explicit user consent form documentations
Range:	
Usage note:	Should be the Hashtag string to report the information about the explicit user consent form documentations.
See also:	n/a

"active":

Semantic attribute:	core:active
Definition:	Boolean to define if the DataOffering is ready to be visible
Range:	
Usage note:	Should be the Boolean to define if the DataOffering is ready to be visible. True or false.
See also:	n/a

"inSharedNetwork":

Semantic attribute:	core:inSharedNetwork
Definition:	Boolean to define if the DataOffering is shared by the marketplace to be visible and consumable by all actors in the i3-MARKET network
Range:	n/a
Usage note:	Should be the Boolean to define if the DataOffering is shared by the marketplace to be visible and consumable by all actors in the i3-MARKET network. True or false.
See also:	n/a

"personalData":

Semantic attribute:	core:personalData
Definition:	Boolean: To define if the data offering offers a dataset that contains personal data
Range:	n/a
Usage note:	Should be the Boolean to define if the data offering offers a dataset that contains personal data.
See also:	n/a

"dataOfferingTitle":

Semantic attribute:	core:dataOfferingTitle
Definition:	The title of the DataOffering
Range:	xsd:string
Usage note:	A name to identify the DataOffering. A few words only that summarize the offering.
See also:	n/a

"dataOfferingDescription":

Semantic attribute:	core:dataOfferingDescription
Definition:	A description of the DataOffering
Range:	xsd:string
Usage note:	Used to have description text to describe what the data offering is about.This can be a long block of text. At least 1000 characters shall be reserved for this.
See also:	n/a

"category":

Semantic attribute:	core:category
Definition:	A category to have a high-level classification of domain for the DataOffering
Range:	xsd:anyURI
Usage note:	Use the categories naming schema defined for high-level categories as URIs:Categories should only be added by extending the categories list. This is done by the community.The parameter should be checked against this list. If it does not match, return an error.

See also:	Categories in table below prefix: dataCatagory <http://i3.market.eu/auth/dataCatagory> dataCatagory:Automotive Data categories (as per definitions in Gitlab file:https://gitlab.com/i3-MARKET/code/data-models/-/blob/master/Version-1/DataOfferingCategory.ttl) <http://i3.market.eu/auth/dataCatagory/Manufacturing> <http://i3.market.eu/auth/dataCatagory/Automotive> <http://i3.market.eu/auth/dataCatagory/Wellbeing> <http://i3.market.eu/auth/dataCatagory/Agriculture> <http://i3.market.eu/auth/dataCatagory/Culture> <http://i3.market.eu/auth/dataCatagory/Economy> <http://i3.market.eu/auth/dataCatagory/Education> <http://i3.market.eu/auth/dataCatagory/Energy> <http://i3.market.eu/auth/dataCatagory/Environment> <http://i3.market.eu/auth/dataCatagory/Government> <http://i3.market.eu/auth/dataCatagory/Health> <http://i3.market.eu/auth/dataCatagory/International> <http://i3.market.eu/auth/dataCatagory/Justice> <http://i3.market.eu/auth/dataCatagory/Regions> <http://i3.market.eu/auth/dataCatagory/Society> <http://i3.market.eu/auth/dataCatagory/Science> <http://i3.market.eu/auth/dataCatagory/Transport> See also file DataOfferingCategory.ttl

"status":

Semantic attribute:	core:status
Definition:	To define the DataOffering status
Range:	xsd:string
Usage note:	Possible values:"Inactive": The offer is not visible but still exists and can be activated again."ToBeDeleted": Data offer is still available and visible but will be deleted once the last contract on this offer expired. No new purchases are allowed on it."Deleted": The offer is not visible and cannot be activated again. It is no longer available for consumers or providers.
Note:	Rename this field to "*Status*". Possible values:"Inactive": The offer is not visible but still exists and can be activated again."ToBeDeleted": Data offer is still available and visible but will be deleted once the last contract on this offer expired. No new purchases are allowed on it."Deleted": The offer is not visible and cannot be activated again. It is no longer available for consumers or providers.

"dataOfferingExpirationTime":

Semantic attribute:	core:dataOfferingExpirationTime
Definition:	Expiration time of DataOffering in case
Range:	Can be: xsd:dateTime
Usage note:	The dateTime data type is used to specify a date and a time. The dateTime is specified in the following form "YYYY-MM-DDThh:mm:ss" where: • YYYY indicates the year • MM indicates the month • DD indicates the day • T indicates the start of the required time section • hh indicates the hour • mm indicates the minute • ss indicates the second **Note:** All components are required! The following is an example of a dateTime declaration in a schema: "2002-05-30T09:00:00".
See also:	n/a

"dataOfferingCreated": (this can be created automatically by the system at registration time, by engine timestamp, instead of manually by market...)

RDF property	core:dataOfferingCreated
Definition:	Date of formal issuance (e.g., publication) of the data offering
Range:	Encoded using the relevant ISO 8601 date and time compliant string (DATETIME) and typed using the appropriate XML schema datatype (XMLSCHEMA11-2) (xsd:dateTime)
Usage note:	This property *should* be set using the first known date of issuance. The date of the initial publication of this data offering in i3-MARKET.
See also:	Property: release date

"lastModified": (this can be created automatically by the system at registration time, by engine timestamp, instead of manually by market...)

Semantic attribute:	core:lastModified
Definition:	Most recent date on which the data offering was changed, updated, or modified

Range:	Encoded using the relevant ISO 8601 date and time compliant string (DATETIME) and typed using the appropriate XML schema datatype (XMLSCHEMA11-2) (xsd:dateTime)
Usage note:	The value of this property indicates a change to the data offering record. An absent value *may* indicate that the item has never changed after itsinitial publication, that the date of last modification is not known, or that the item is continuously updated.
See also:	Property: frequency, Property: update/modification date, and Property: update/modification date in DCAT 3 webpage

"versionNotes":

Semantic attribute:	adms:versionNotes
Definition:	A description of changes between this version and the previous version of the resource (VOCAB-ADMS)
Range:	xsd:string
Usage note:	In case of backward compatibility issues with the previous version of the resource, a textual description of them *should* be specified by using this property.
See also:	Property: current version, Property: has version, Property: is replaced by, Property: is version of, Property: previous version, Property: release date, Property: replaces, Property: status, and Property: version notes

"previousVersion":

Semantic attribute:	dcat:previousVersion
Definition:	The previous version of a resource in a lineage (PAV)
Range:	xsd:anyURI
Usage note:	This property is meant to be used to specify a version chain, consisting of snapshots of a resource. The notion of version used by this property is limited to versions resulting from revisions occurring to a resource as part of its lifecycle. One of the typical cases here is representing the history of the versions of a dataset that have been released over time.
See also:	Property: current version, Property: has version, Property: is replaced by, Property: is version of, Property: previous version, Property: release date, Property: replaces, Property: status, and Property: version notes

"replaces":

Semantic attribute:	dcterms:replaces
Definition:	A related resource that is supplanted, displaced, or superseded by the described resource (DCTERMS)
Range:	xsd:anyURI
Usage note:	Resource replaced.
See also:	Property: current version, Property: has version, Property: is replaced by, Property: is version of, Property: previous version, Property: release date, Property: replaces, Property: status, and Property: version notes

"previousVersion":

Semantic attribute:	dcat:previousVersion
Definition:	The previous version of a resource in a lineage (PAV)
Range:	xsd:anyURI
Usage note:	This property is meant to be used to specify a version chain, consisting of snapshots of a resource. The notion of version used by this property is limited to versions resulting from revisions occurring to a resource as part of its lifecycle. One of the typical cases here is representing the history of the versions of a dataset that have been released over time.
See also:	Property: current version, Property: has version, Property: is replaced by, Property: is version of, Property: previous version, Property: release date, Property: replaces, Property: status, and Property: version notes

"contractParameters":
   {
   "interestOfProvider":

Semantic attribute:	core:interestOfProvider
Definition:	This property is used to identify the interest of the data owner/provider related to the trading/sharing of their data assets. The following possibilities exist: • Free sharing • Quotation • Selling of data (e.g., just earning money by selling the data, no specific feedback on these data by a data consumer expected)
Range:	xsd:string
Usage note:	It could be simple notations like: Free Sharing – Quotation – Selling of data; or we can decide to have specific definitions for our system.
See also:	n/a

"interestDescription":

Semantic attribute:	core:interestDescription
Definition:	Data provider can specify which sort of quotation he wants exactly, e.g., quotation for maintenance service or quotation for optimization of production
Range:	xsd:string
Usage note:	More text description of the interest of the data owner/provider related to the trading/sharing of their data assets.Example: "This data is shared only for the purpose of creating a quotation for maintenance for the production machines described in the dataset. Any other use of this data is not permitted".
Note:	n/a

"hasGoverningJurisdiction":

Semantic attribute:	core:hasGoverningJurisdiction
Definition:	The file format of the distribution
Range:	xsd:string (or xsd:anyURI)
Usage note:	Can be string naming like: GLOBAL US-JURISDICTION EU-JURISDICTION (or we use URIs to define the specific terms for jurisdictions) To be extended to define a list of jurisdictions that are valid here.
See also:	n/a

"purpose":

Semantic attribute:	core:purpose
Definition:	Purpose of the agreement
Range:	xsd:string
Usage note:	Short label for the purpose. In case we could have specific terminology for define list of @purpose@ terms.
Note:	This parameter is part of the contractual parameters. Ask contract partners, what this is for (Susanne).

"purposeDescription":

Semantic attribute:	core:purposeDescription
Definition:	In case full text description of describing the reasons behind the creation of the agreement
Range:	xsd:string
Usage note:	Text description.
Note:	This parameter is part of the contractual parameters. Ask contract partners, what this is for (Susanne).

"hasIntendedUse":
{
"processData": "true OR false"

Semantic attribute:	core:processData
Definition:	If consumer allowed to process data
Range:	xsd:boolean
Usage note:	The value space of xsd:boolean is true and false. Its lexical space accepts true, false, "TRUE", or "FALSE".
Note:	Part of contractual parameters. Ask contract partners, what this is for.Make this parameter to type Boolean.

"shareDataWithThirdParty": "true OR false"

Semantic attribute:	core:shareDataWithThirdParty
Definition:	If consumer allowed to share data with third parties
Range:	xsd:boolean
Usage note:	The value space of xsd:boolean is true and false. Its lexical space accepts true, false, "TRUE", or "FALSE".
Note:	Part of contractual parameters. Ask contract partners, what this is for.Make this parameter to type Boolean.

"editData": "true OR false"

Semantic attribute:	core:editData
Definition:	If consumer is allowed to edit the data
Range:	xsd:boolean
Usage note:	The value space of xsd:boolean is true and false. Its lexical space accepts true, false, "TRUE", or "FALSE".
Note:	Part of contractual parameters. Ask contract partners, what this is for.Make this parameter to type Boolean.

},

"hasLicenseGrant":
  {
  "paidUp": "true OR false"

Semantic attribute:	core:paidUp
Definition:	If licence grant to paidUp
Range:	xsd:boolean
Usage note:	The value space of xsd:boolean is true and false. Its lexical space accepts true, false, "TRUE", or "FALSE".
Note:	Part of contractual parameters. Ask contract partners, what this is for.

"transferable": "true OR false"

Semantic attribute:	core:transferable
Definition:	If licence is transferable
Range:	xsd:boolean
Usage note:	The value space of xsd:boolean is true and false. Its lexical space accepts true, false, "TRUE", or "FALSE".
See also:	n/a

"exclusiveness": "true OR false"

Semantic attribute:	core:exclusiveness
Definition:	If licence grant exclusiveness
Range:	xsd:boolean
Usage note:	The value space of xsd:boolean is true and false. Its lexical space accepts true, false, "TRUE", or "FALSE".
See also:	n/a

"revocable": "true OR false"

Semantic attribute:	core:revocable
Definition:	If licence is revocable
Range:	xsd:boolean
Usage note:	The value space of xsd:boolean is true and false. Its lexical space accepts true, false, "TRUE", or "FALSE".
See also:	n/a

"processing": "true OR false"

Semantic attribute:	core:processing
Definition:	If licence grant data to be processed
Range:	xsd:boolean
Usage note:	The value space of xsd:boolean is true and false. Its lexical space accepts true, false, "TRUE", or "FALSE".
Note:	Part of contractual parameters. Ask contract partners, what this is for.

"modifying": "true OR false"

Semantic attribute:	core:modifying
Definition:	If licence grant data to be modified
Range:	xsd:boolean
Usage note:	The value space of xsd:boolean is true and false. Its lexical space accepts true, false, "TRUE", or "FALSE".
Note:	Part of contractual parameters. Ask contract partners, what this is for.

"analyzing": "true OR false"

Semantic attribute:	core:analyzing
Definition:	If licence grant data to be analysed
Range:	xsd:boolean
Usage note:	The value space of xsd:boolean is true and false. Its lexical space accepts true, false, "TRUE", or "FALSE".
Note:	Part of contractual parameters. Ask contract partners, what this is for.

"storingData": "true OR false"

Semantic attribute:	core:storingData
Definition:	If licence grant to store data
Range:	xsd:boolean
Usage note:	The value space of xsd:boolean is true and false. Its lexical space accepts true, false, "TRUE", or "FALSE".
Note:	Part of contractual parameters. Ask contract partners, what this is for.

"storingCopy": "true OR false"

Semantic attribute:	core:storingCopy
Definition:	If licence grant to store a copy data
Range:	xsd:boolean
Usage note:	The value space of xsd:boolean is true and false. Its lexical space accepts true, false, "TRUE", or "FALSE".
Note:	Part of contractual parameters. Ask contract partners, what this is for.

"reproducing": "true OR false"

Semantic attribute:	core:reproducing
Definition:	If licence grant to reproduce data
Range:	xsd:boolean
Usage note:	The value space of xsd:boolean is true and false. Its lexical space accepts true, false, "TRUE", or "FALSE".
Note:	Part of contractual parameters. Ask contract partners, what this is for.

"distributing": "true OR false"

Semantic attribute:	core:distributing
Definition:	If licence grant to distribute data
Range:	xsd:boolean
Usage note:	The value space of xsd:boolean is true and false. Its lexical space accepts true, false, "TRUE", or "FALSE".
Note:	Part of contractual parameters. Ask contract partners, what this is for.

"loaning": "true OR false"

Semantic attribute:	core:loaning
Definition:	If licence grant to loan data
Range:	xsd:boolean
Usage note:	The value space of xsd:boolean is true and false. Its lexical space accepts true, false, "TRUE", or "FALSE".
Note:	Part of contractual parameters. Ask contract partners, what this is for.

"selling": "true OR false"

Semantic attribute:	core:selling
Definition:	If licence grant to sell data
Range:	xsd:boolean
Usage note:	The value space of xsd:boolean is true and false. Its lexical space accepts true, false, "TRUE", or "FALSE".
Note:	Part of contractual parameters. Ask contract partners, what this is for.

"renting": "true OR false"

Semantic attribute:	core:renting
Definition:	If licence grant to rent data
Range:	xsd:boolean
Usage note:	The value space of xsd:boolean is true and false. Its lexical space accepts true, false, "TRUE", or "FALSE".
Note:	Part of contractual parameters. Ask contract partners, what this is for.

"furtherLicensing": "true OR false"

Semantic attribute:	core:furtherLicensing
Definition:	If licence grant for further licensing
Range:	xsd:boolean
Usage note:	The value space of xsd:boolean is true and false. Its lexical space accepts true, false, "TRUE", or "FALSE".
Note:	Part of contractual parameters. Ask contract partners, what this is for.

"leasing": "true OR false"

Semantic attribute:	core:leasing
Definition:	If licence grant to lease data
Range:	xsd:boolean
Usage note:	The value space of xsd:boolean is true and false. Its lexical space accepts true, false, "TRUE", or "FALSE".
Note:	Part of contractual parameters. Ask contract partners, what this is for.

```
}     } ,
```

"hasPricingModel":
  {
   "pricingModelName":

Semantic attribute:	pricingmodel:pricingModelName
Definition:	The name to define the legacy, by marketplace, pricing model related to the data offering
Range:	xsd:string
Usage note:	Pricing models are individually defined by marketplaces. No common pricing model will be defined for i3-MARKET. Maybe try to generalize existing pricing models.
See also:	

"basicPrice":

Semantic attribute:	pricingmodel:basicPrice
Definition:	The generic basic price for the traded data for basic cost of trade
Range:	xsd:double
Usage note:	Number related to price.
See also:	

"currency":

Semantic attribute:	pricingmodel:currency
Definition:	The file format of the distribution
Range:	xsd:string
Usage note:	Using ISO 4215 currency terminology.
See also:	lis-ISO-4217-Currencyt_one.xml See XML file for three-letter abbreviations. lis-ISO-4217-Currencyt_one.xml

"hasPaymentOnSubscription":
  {
   "timeDuration":

Semantic attribute:	pricingmodel:timeDuration
Definition:	Time duration of subscription
Range:	xsd:anyURI

Usage note:	Or generic xsd:string text with labels for duration vocabulary or URIs with vocabulary like: "http://reference.data.gov.uk/def/intervals/Day" "http://reference.data.gov.uk/def/intervals/Hour" "http://reference.data.gov.uk/def/intervals/Minute" "http://reference.data.gov.uk/def/intervals/Month" "http://reference.data.gov.uk/def/intervals/Quarter" "http://reference.data.gov.uk/def/intervals/Second" Price is per timeDuration. For example, if parameter is "Second" here, then the specified price is per second (€/ s).
See also:	Terms in intervals.rdf

"description":

Semantic attribute:	dcterms:description
Definition:	The description of payment on subscription
Range:	xsd:string
Usage note:	Text description.
See also:	n/a

"repeat":

Semantic attribute:	pricingmodel:repeat
Definition:	If subscription can be repeated define the frequency, e.g., daily, monthly, etc.
Range:	xsd:anyURI
Usage note:	We can use specific vocabulary For example, in freq.ttl definitions like: http://purl.org/cld/freq/daily freq:monthly freq:weekly ....
See also:	See also freq.ttl or frequency.ttl.txt

"hasSubscriptionPrice":

Semantic attribute:	pricingmodel:hasSubscriptionPrice
Definition:	Price allocated to subscription payment type
Range:	xsd:double
Usage note:	Price.
See also:	n/a

}   ,

"hasPaymentOnPlan":
{
There may be things like basic plan, premium plans, etc. that gives access to certain types of data. Which are difficult to implement in i3-MARKET.

Example for other usage: Deliver data only once a month or once every *x* period.Optional parameter does not have to be used.

"description":

Semantic attribute:	pricingmodel:planDescription
Definition:	The text description of plan
Range:	Xsd:string
Usage note:	Description text.
See also:	n/a

"planDuration":

Semantic attribute:	pricingmodel:planDuration
Definition:	The duration of the plan
Range:	xsd:anyURI
Usage note:	Or generic xsd:string text with labels for duration vocabulary or URIs with vocabulary like: "http://reference.data.gov.uk/def/intervals/Day" "http://reference.data.gov.uk/def/intervals/Hour" "http://reference.data.gov.uk/def/intervals/Minute" "http://reference.data.gov.uk/def/intervals/Month" "http://reference.data.gov.uk/def/intervals/Quarter" "http://reference.data.gov.uk/def/intervals/Second"
See also:	Terms in intervals.rdf

"hasPlanPrice": "string"

Semantic attribute:	pricingmodel:hasPlanPrice
Definition:	The price of the plan
Range:	xsd:double
Usage note:	Price.
See also:	n/a

}   ,

"hasPaymentOnApi":
    {
        "description":

Semantic attribute:	Dcterms:description
Definition:	The text description of payment type
Range:	Xsd:string
Usage note:	Description text.
Note:	Optional. Useful for Agora.

        "numberOfObject":

Semantic attribute:	pricingmodel:numberObject
Definition:	Number of objects for API handle payments
Range:	Xsd:double
Usage note:	
Note:	Optional. Useful for Agora.

        "hasAPIPrice": "string"

Semantic attribute:	pricingmodel:hasAPIPrice
Definition:	The price of the API payment type
Range:	xsd:double
Usage note:	Price.
Note:	Optional. Useful for Agora.

    } ,
"hasPaymentOnUnit":
    {
        "description":

Semantic attribute:	Dcterms:description
Definition:	The text description of payment type
Range:	Xsd:string
Usage note:	Description text. Purchase a cluster of data. Sets of data. One cluster is a group of datasets.
See also:	n/a

"dataUnit":

Semantic attribute:	pricingmodel:dataUnit
Definition:	Data unit type handle by service
Range:	Xsd:string
Usage note:	Define what the unit resembles. Example: A predefined dataset. A "Unit" of transaction as indicated in specification of the service method of exchange.
See also:	Data unit type – In telecommunications, a *protocol data unit (PDU)* is a single unit of information transmitted among peer entities of a computer network. For example, the data unit in which data are packeted when transmitted in streams. Also, e.g., a data unit that contains one or many stream data objects.

"hasUnitPrice": "string"

Semantic attribute:	pricingmodel:hasUnitPrice
Definition:	The price of the unit by payment type
Range:	xsd:double
Usage note:	Price per data unit.
See also:	n/a

} ,

"hasPaymentOnSize":
{
"description":

Semantic attribute:	Dcterms:description
Definition:	The text description of payment type
Range:	Xsd:string
Usage note:	Description text.
See also:	n/a

"dataSize":

Semantic attribute:	pricingmodel:dataSize
Definition:	The size of data exchanged for payment
Range:	Typically typed as xsd:nonNegativeInteger
Usage note:	The size in bytes can be approximated (as a non-negative integer) when the precise size is not known. While it is recommended that the size be given as an integer, alternative literals such as "1.5 MB" are sometimes used.
See also:	We can decide to use a specific vocabulary

"hasSizePrice": "string"

Semantic attribute:	pricingmodel:hasSizetPrice
Definition:	The price of the unit by payment type
Range:	xsd:double
Usage note:	Price, e.g., pay per megabyte of data.
See also:	n/a

} ,

"hasFreePrice":
{
"hasPriceFree": "FREE"

Semantic attribute:	pricingmodel:hasPriceFree
Definition:	The data is shared for free
Range:	Xsd:string
Usage note:	"FREE". Data is for free, no payment needed.
See also:	We might use an URI as Pricingmodel:Free as unique term

} } ,

"hasDataset":
{ (Dataset description)
Description of the datasets contained. Note: This is not a description of the individual data items but an overview.

"title":

Semantic attribute:	dcterms:title
Definition:	A name given to the dataset
Range:	Xsd:string [rdfs:Literal]
Usage note:	Title.
See also:	n/a

"keyword":

Semantic attribute:	dcat:keyword
Definition:	A keyword or tag describing the resource
Range:	Xsd:string [rdfs:Literal]
Usage note:	Text keywords (in case we can decide to have a selection of terminologies to set as keywords). One or more keywords describing the data.
See also:	To have multiple keywords, you can have multiple instances of the property "keyword"

"description":

Semantic attribute:	dcterms:description
Definition:	A free-text account of the dataset
Range:	Xsd:string [rdfs:Literal]
Usage note:	Description text of dataset.
See also:	n/a

"issued":

RDF property	dcterms:issued
Definition:	Date of formal issuance (e.g., publication) of the distribution
Range:	Encoded using the relevant ISO 8601 date and time compliant string (DATETIME) and typed using the appropriate XML schema datatype (XMLSCHEMA11-2) (xsd:dateTime)
Usage note:	This property *should* be set using the first known date of issuance. The date of the initial publication of this dataset in i3-MARKET.
See also:	§ 6.4.7 Property: release date

"modified":

Semantic attribute:	dcterms:modified
Definition:	Most recent date on which the item was changed, updated, or modified
Range:	Encoded using the relevant ISO 8601 date and time compliant string (DATETIME) and typed using the appropriate XML schema datatype (XMLSCHEMA11-2) (xsd:dateTime)
Usage note:	The value of this property indicates a change to the actual item, not a change to the catalogue record. An absent value *may* indicate that the item has never changed after itsinitial publication, that the date of last modification is not known, or that the item is continuously updated.
See also:	§ 6.6.2 Property: frequency, § 6.5.4 Property: update/modification date, and § 6.8.4 Property: update/modification date in DCAT 3 webpage

"temporal":

Semantic attribute:	dcterms:temporal
Definition:	The temporal period that the dataset covers
Range:	In general, used singularly can be used URIs as in intervals vocab OR dcterms:PeriodOfTime (an interval of time that is named or defined by its start and end dates)

Usage note:	In case we extend the model to serve the temporal coverage of a dataset may be encoded as an instance of dcterms:PeriodOfTime, or may be indicated using an IRI reference (link) to a resource describing a time period or interval. For example, as [a dcterms:PeriodOfTime] dcat:startDate "2016-03-04"∧∧xsd:dateTime; dcat:endDate "2018-08-05"∧∧xsd:dateTime;
See also:	Intervals.rdf

"language":

Semantic attribute:	dcterms:language
Definition:	A language of the item. This refers to the natural language used for textual metadata (i.e., titles, descriptions, etc.) of a catalogued resource (i.e., dataset or service) or the textual values of a dataset distribution
Range:	Resources defined by the Library of Congress (ISO 639-1, ISO 639-2) *should* be used If an ISO 639-1 (two-letter) code is defined for language, then its corresponding IRI *should* be used; if no ISO 639-1 code is defined, then IRI corresponding to the ISO 639-2 (three-letter) code *should* be used
Usage note:	Repeat this property if the resource is available in multiple languages.
See also:	Also if representations of a dataset are available for each language separately, define an instance of dcat:Distribution for each language and describe the specific language of each distribution using dcterms:language (i.e., the dataset will have multiple dcterms:language values and each distribution will have just one as the value of its dcterms:language property).

"spatial":

Semantic attribute:	dcterms:spatial
Definition:	The geographical area covered by the dataset
Range:	Xsd:anyURI to use in case using a IRI reference (link) to a resource describing a location. It is recommended that links are to entries in a well-maintained gazetteer such as Geonames Or a dcterms:Location (a spatial region or named place)
Usage note:	The spatial coverage of a dataset may be encoded as an instance of dcterms:Location. Or may be indicated using an IRI reference (link) to a resource describing a location. It is recommended that links are to entries in a well-maintained gazetteer such as Geonames.

See also:	For example, for bbox dcterms:spatial [[a dcterms:Location] dcat:bbox """POLYGON[[3.053 47.975 , 7.24 47.975 , 7.24 53.504 , 3.053 53.504 , 3.053 47.975]]""" ;]

"accrualPeriodicity":

Semantic attribute:	dcterms:accrualPeriodicity
Definition:	The frequency at which a dataset is published
Range:	xsd:anyURI
Usage note:	We can use specific vocabulary For example, in freq.ttl definitions like: http://purl.org/cld/freq/daily freq:monthly freq:weekly.
See also:	See also freq.ttl or at frequency.ttl.txt

"temporalResolution":

Semantic attribute:	dcat:temporalResolution
Definition:	Minimum time period resolvable in the dataset
Range:	xsd:duration
Usage note:	If the dataset is a time-series, this should correspond to the spacing of items in the series. For other kinds of dataset, this property will usually indicate the smallest time difference between items in the dataset.
See also:	n/a

"theme": [

Semantic attribute:	dcat:theme
Definition:	A (sub-)category of the resource. A resource can have multiple themes
Range:	It would be better to have xsd:anyURI with URIs that represent the various terms in a vocabulary (to be defined with pilot partners for terms related to domains)
Usage note:	Use this for domain-specific categories. For example, subcategories like production machines, assembly lines, etc.To be defined by each application domain.Theme can be used multiple times to provide multiple subcategories. The set of themes used to categorize the resources are organized in a skos:ConceptScheme, skos:Collection, owl:Ontology, or similar, describing all the categories and their relations in the catalogue.
See also:	

],

"distribution": (Distribution: A specific representation of a dataset. A dataset might be available in multiple serializations that may differ in various ways, including natural language, media-type or format, schematic organization, temporal and spatial resolution, and level of detail or profiles [which might specify any or all of the above]).

{

"title":

Semantic attribute:	dcterms:title
Definition:	A name given to the distribution
Range:	Xsd:string [rdfs:Literal]
Usage note:	Title.
See also:	n/a

"description":

Semantic attribute:	dcterms:description
Definition:	A free-text account of the distribution
Range:	Xsd:string [rdfs:Literal]
Usage note:	Description text of dataset.
See also:	n/a

"license":

Semantic attribute:	dcterms:license
Definition:	A legal document under which the distribution is made available
Range:	dcterms:LicenseDocument
Usage note:	For interoperability, it is recommended to use canonical IRIs of well-known licenses such as those defined by Creative Commons. Information about licenses and rights *should* be provided on the level of distribution. Information about licenses and rights *may* be provided for a dataset in addition to but not instead of the information provided for the distributions of that dataset. Providing license or rights information for a dataset that is different from information provided for a distribution of that dataset *should* be avoided as this can create legal conflicts. See also guidance at §9. License and rights statements.
See also:	Property: rights Property: license **ToDo: Describe a list of possible licenses here.**

"accessRights":

Semantic attribute:	dcterms:accessRights
Definition:	Information about who can access the resource or an indication of its security status
Range:	dcterms:LicenseDocument
Usage note:	Information about licenses and rights *may* be provided for the resource. To express statements concerning only access rights (e.g., whether data can be accessed by anyone or just by authorized parties). Access rights can also be expressed as code lists/taxonomies. Examples include the access rights code list (EUV-AR) used in (DCAT-AP) and the Eprints Access Rights Vocabulary Encoding Scheme.
See also:	Property: rights dcterms:accessRights <http://publications.europa.eu/resource/authority/access-right/PUBLIC> ; dcterms:conformsTo <http://www.opengis.net/def/serviceType/ogc/csw> ;

"downloadType":

Semantic attribute:	core:downloadType
Definition:	Information about download type (it means "Stream" or "Bulk" dataset download)
Range:	xsd:string
Usage note:	To use a set of words like "Stream" and "Bulk".
See also:	n/a

"dataStream":

Semantic attribute:	core:dataStream
Definition:	Boolean attribute to check if the dataset is offered as a stream or not
Range:	
Usage note:	Should be the Boolean attribute to check if the dataset is offered as stream or not in the "Distribution" class block.
See also:	n/a

"conformsTo":

Semantic attribute:	dcterms:conformsTo
Definition:	An established standard to which the distribution conforms (very optional)
Range:	dcterms:Standard (A basis for comparison; a reference point against which other things can be evaluated.)
Usage note:	This property *should* be used to indicate the model, schema, ontology, view, or profile that this representation of a dataset conforms to. This is (generally) a complementary concern to the media-type or format. This is a link to a specific file that describes the data in a domain specific format. It can also be a text in a freely definable format.
See also:	Property: format, Property: media type Also check file-type.ttl.txt

"mediaType":

Semantic attribute:	dcat:mediaType
Definition:	The media-type of the distribution as defined by IANA (IANA-MEDIA-TYPES)
Range:	Xsd:anyURI [dcterms:MediaType]
Usage note:	dcat:mediaType *should* be used if the type of the distribution is defined by IANA (IANA-MEDIA-TYPES). https://www.iana.org/assignments/media-types/ For example, mediaType <http://www.iana.org/assignments/media-types/application/ld+json> For example, a link to a XML, csv, or JSON file, to describe the data format.
See also:	Property: media type, Property: conforms to Check also file-type.ttl.txt

"packageFormat":

Semantic attribute:	dcat:packageFormat
Definition:	The package format of the distribution in which one or more data files are grouped together, e.g., to enable a set of related files to be downloaded together
Range:	Xsd:anyURI [dcterms:MediaType]
Usage note:	In case it is compressed, this could be .zip, .rar, etc. This property to be used when the files in the distribution are packaged, e.g., in a TAR file, a Frictionless Data Package, or a Bagit file. Theformat *should* be expressed using a media-type as defined by IANA (IANA-MEDIA-TYPES), if available.
See also:	Property: compression format.

"accessService": (info inside distribution for service that serves the distributions of the datasets)
{
"conformsTo":

Semantic attribute:	dcterms:conformsTo
Definition:	An established standard to which the distribution conforms
Range:	dcterms:Standard (A basis for comparison; a reference point against which other things can be evaluated.)
Usage note:	This property *should* be used to indicate the model, schema, ontology, view, or profile that this representation of a dataset conforms to. This is (generally) a complementary concern to the media-type or format.
See also:	Property: conforms to

"endpointDescription":

Semantic attribute:	dcat:endpointDescription
Definition:	A description of the services available via the endpoints, including their operations, parameters, etc.
Range:	xsd:string
Usage note:	The endpoint description gives specific details of the actual endpoint instances, while dcterms:conformsTo is used to indicate the general standard or specification that the endpoints implement. An endpoint description may be expressed in a machine-readable form, such as an OpenAPI (Swagger) description (OpenAPI), an OGC GetCapabilities response (WFS), (ISO-19142), (WMS), (ISO-19128), a SPARQL service description (SPARQL11-SERVICE-DESCRIPTION), an (OpenSearch) or (WSDL20) document, a Hydra API description (HYDRA), and else in text or some other informal modes if a formal representation is not possible.
See also:	n/a

"endpointURL":

Semantic attribute:	dcat:endpointURL
Definition:	The root location or primary endpoint of the service (a Web-resolvable IRI)
Range:	xsd:anyURI
Usage note:	The URL address of the resource via service.
See also:	n/a

"servesDataset":

Semantic attribute:	dcat:servesDataset
Definition:	A collection of data that this data service can distribute. The dataset ID or name and files
Range:	xsd:string
Usage note:	To point to the datasets that are served via the data service.
See also:	n/a

"serviceSpecs": "string"

Semantic attribute:	core:serviceSpecs
Definition:	Description of service specification for more details on the data service implementations
Range:	
Usage note:	To extend in case the description of data service to add more detailed descriptions on the system.To describe more details about the service, e.g. QoS, etc.
See also:	n/a

"dataExchangeSpec": (info inside accessService block for data exchange specifications that serve the distributions of the datasets)
{
"encAlg": "string"

Semantic attribute:	core:encAlg
Definition:	Encryption algorithm used to encrypt blocks. Either AES-128-GCM ('A128GCM') or AES-256-GCM ('A256GCM)
Range:	
Usage note:	Encryption algorithm used to encrypt blocks. Either AES-128-GCM ('A128GCM') or AES-256-GCM ('A256GCM).
See also:	n/a

"signingAlg": "string"

Semantic attribute:	core:signingAlg
Definition:	Signing algorithm used to sign the proofs. Like ECDSA secp256r1 with key lengths: either "ES256", "ES384", or "ES512"
Range:	n/a
Usage note:	Signing algorithm used to sign the proofs. It is ECDSA secp256r1 with key lengths: either "ES256", "ES384", or "ES512".
See also:	n/a

"hashAlg": "string"

Semantic attribute:	core:hashAlg
Definition:	Hash algorithm used to compute digest/commitments. It is SHA2 with different output lengths: either "SHA-256", "SHA-384", or "SHA-512"
Range:	
Usage note:	Hash algorithm used to compute digest/commitments. It is SHA2 with different output lengths: either "SHA-256", "SHA-384", or "SHA-512".
See also:	n/a

"ledgerContractAddress": "string"

Semantic attribute:	core:ledgerContractAddress
Definition:	The ledger smart contract address (hexadecimal) on the DLT
Range:	n/a
Usage note:	The ledger smart contract address (hexadecimal) on the DLT.
See also:	n/a

"ledgerSignerAddress": "string"

Semantic attribute:	core:ledgerSignerAddress
Definition:	The orig (data provider) address in the DLT (hexadecimal)
Range:	n/a
Usage note:	The orig (data provider) address in the DLT (hexadecimal).
See also:	n/a

"pooToPorDelay": "number"

Semantic attribute:	core:pooToPorDelay
Definition:	Maximum acceptable delay between the issuance of the proof of origin (PoO) by the orig and the reception of the proof of reception (PoR) by the orig
Range:	n/a
Usage note:	Maximum acceptable delay between the issuance of the proof of origin (PoO) by the orig and the reception of the proof of reception (PoR) by the orig.
See also:	n/a

"pooToPopDelay": "number"

Semantic attribute:	core:pooToPopDelay
Definition:	Maximum acceptable delay between the issuance of the proof of origin (PoP) by the orig and the reception of the proof of publication (PoR) by the dest
Range:	
Usage note:	Maximum acceptable delay between the issuance of the proof of origin (PoP) by the orig and the reception of the proof of publication (PoR) by the dest.
See also:	n/a

"pooToSecretDelay": "number"

Semantic attribute:	core:pooToSecretDelay
Definition:	If the dest (data consumer) does not receive the PoP, it could still get the decryption secret from the DLT. This defines the maximum acceptable delay between the issuance of the proof of origin (PoP) by the orig and the publication (block time) of the secret on the blockchain
Range:	n/a
Usage note:	If the dest (data consumer) does not receive the PoP, it could still get the decryption secret from the DLT. This defines the maximum acceptable delay between the issuance of the proof of origin (PoO) by the orig and the publication (block time) of the secret on the blockchain.
See also:	n/a

```
        }      }
}    ],
```

"datasetInformation": (a description of types that represent attributes of observations, measurements, fields, etc., in the dataset to describe the information and structure of the raw real data in the datasets)

{

"measurementType":

Semantic attribute:	core:measurementType
Definition:	The data types that represent attributes of observations and measurements in the dataset
Range:	xsd:anyURI
Usage note:	Simple text strings or the use of specific vocabularies collected to support domains For example, like the vocabulary created for wellbeing. For example, <http://www.i3-MARKET.eu/wellbeing_annotat ions/Sleep_count_micro_awakenings>. Specific types of measurements for a certain domain. Parameter can be put multiple times in the API call.
See also:	See also example for Wellbeing in DataRecords_Annotations_f or_Wellbeing_datasets_measurements_02.ttl attached to this page but also in gitlab https://gitlab.com/i3-MARKET/code/da ta-models/-/blob/master/Version-1/DataRecords_Annotations_f or_Wellbeing_datasets_measurements_02.ttl

"measurementChannelType":

Semantic attribute:	core:measurementChannelType
Definition:	The data measurement channel types in the dataset
Range:	xsd>string or xsd>anyURI
Usage note:	Simple text strings or the use of specific vocabularies collected to support domains.
See also:	n/a

"sensorId":

Semantic attribute:	core>sensorID
Definition:	Sensor ID
Range:	xsd>string
Usage note:	ID used to identify the sensors in original datasets source.
See also:	n/a

"deviceId":

Semantic attribute:	core>deviceID
Definition:	Device ID
Range:	xsd>string
Usage note:	ID used to identify the devices in original datasets source.
See also:	n/a

"cppType":

Semantic attribute:	core:cppType
Definition:	The cpp types in the dataset. Derived from AGORA requirements
Range:	xsd>string or xsd>anyURI
Usage note:	Simple text strings or the use of specific vocabularies collected to support domains.
See also:	n/a

"sensorType": "string"

Semantic attribute:	core:sensorType
Definition:	The cpp types in the dataset. Derived from wellbeing and AGORA requirements
Range:	xsd>string or xsd>anyURI
Usage note:	Simple text strings or the use of specific vocabularies collected to support domains.
See also:	n/a

4.7 Extended Version of Structure for DatasetInformation

The DatasetInformation module may be extended to add more specific description and structure information related to the raw original data that is contained in the assets/datasets delivered by the providers. This way, users can check and have a better understanding of the underlying data model and associated metadata that describe the data that are transferred. The proposed data model description of the source of the data, each observation and data item in the data with their details like quantity type, data type, unit, resolution, range, etc.

```
datasetInformation": [
    {
      "dataSource":
        {
          "sourceDescription": "string"
          "sourceLocation": "string",
          "sourceCountry": "string",        (location
    info also in general in higher level of "Dataset"
    block description)
              "generatedBy" {
```

```
                    "platform": "string",
                    "device": "string",
                    "sensor": "string",
                    "sensorResolution": "string",
                    "procedure": "string"
                }
            }
        "document": {
                "name":      "string",
                "documentDescription":      "string",
                "comment": "string",
                "timeResolution":      "string"
        }
        "observation" (dataItem): [
            {
                "observationNname":      "string",
                "observationDdescription":      "string",
                "observedProperty":      "string",
                "dataType":      "string",
                "accuracy":      "string",
                "timeResolution":      "string",
                "measure": {
                            "quantityKind":
    "string",
                            "unit":      "string",
                            "value":      "string"
                }
                "range":
                {
                    "rangeType": "string",
                    "min": "string",
                    "max": "string"
                }
                "subObservation" (subDataItem) [
                {
                        "observationNname":      "string",
                        "observationDdescription":
    "string",
                        "observedProperty":      "string",
                        "dataType":      "string",
                        "accuracy":      "string",
                        "timeResolution":      "string",
                        "measure": {
                            "quantityKind":
    "string",
                            "unit":      "string",
                            "value":      "string",
                }
                        "range":
                {
                    "rangeType": "string",
                    "min": "string",
                    "max": "string",
                }    }
                ]
        } ]
        } ]
```

For the definition and semantic annotations related to quantities and units, we can refer to:

- The Ontology of Units of Measure (OM) 2.0 models concepts and relations are focus on units, quantities, measurements, and dimensions. http://www.ontology-of-units-of-measure.org/page/om-2
- The Quantities, Units, Dimensions and Data Types Ontologies http://www.qudt.org/2.1/catalog/qudt-catalog.html
  - https://qudt.org/
  - https://www.qudt.org/2.1/catalog/qudt-catalog.html
  - https://www.qudt.org/doc/DOC_SCHEMA-DATATYPES.html
  - https://www.qudt.org/doc/DOC_VOCAB-UNITS.html
  - https://www.qudt.org/doc/DOC_VOCAB-QUANTITY-KINDS.html

5

Distributed Data Storage System Considerations

5.1 Objectives

Every federated information system requires means to store and share data securely. The i3-MARKET network is not an exception; hence, a well-thought solution that is secure, reliable, and usable by all entities in the i3-MARKET network is needed. The aim of data storage is to store common data in a federated network of data marketplaces. The common data shared between participating data marketplace instances may include identity information, shared semantic models, meta-information about datasets and offerings, semantic queries, sample data, smart contract templates and instances, crypto tokens, and payments. No single party should fully control the data storage system and there shall be no single point of failure. In order to fulfil the needs of the aforementioned data types, two separate storage solutions are used: the decentralized and the distributed one.

The former supports the management of distributed identities and smart contracts. However, the latter has an important role in data synchronization between different i3-MARKET nodes and, optionally, storage of datasets on sale. Moreover, the distributed storage supports non-repudiation service and auditable accounting features of i3-MARKET.

The design of the distributed storage has been an iterative process.

Data storage system takes full advantage of available base technologies and builds on top of these in order to satisfy i3-MARKET needs and requirements, with a focus on federated system architecture. The underlying technologies chosen for decentralized and distributed storage means are Hyperledger BESU and CockroachDB, respectively.

The federated query engine index (SEED Index) management solution is available and integrated into the i3-MARKET network, deployed as a smart contract on Hyperledger BESU. Moreover, a solution called verifiable data integrity has been implemented on top of auditable accounting to further increase the reliability of data. And, finally, access management solution governing the data access has been designed and implemented, depending on reliable and secure key management solution.

The common data shared between participating data marketplace instances may include identity information, shared semantic models, meta-information about datasets and offerings, semantic queries, sample data, smart contract templates and instances, crypto tokens, and payments. No single party should fully control the data storage system and there shall be no single point of failure.

The high-level capabilities that the data storage aims to provide are:

1. Decentralized storage
2. Distributed storage

The decentralized storage shall provide the highest available security guarantees in a federated network. The decentralized storage subsystem is built on a secure Byzantine fault-tolerant consensus-based distributed ledger. Due to high security requirements, the performance and storage space of such a system may be relatively limited compared to conventional databases.

The distributed storage shall provide a database-like subsystem that is scalable, deployed on all i3-MARKET nodes, has a rich query interface (SQL), and can handle large amounts of data, while the i3-MARKET shall rely on the API of the decentralized storage provided out-of-box.

5.2 Solution Design/Blocks

The storage system consists of two main subsystems for implementing the decentralized storage and distributed storage features, respectively. The subsystems are relatively independent of other systems and also with each other.

The diagram of a decentralized storage subsystem is shown in Figure 5.1. The decentralized storage subsystem is implemented as a blockchain-based distributed ledger network. The software implementation is Hyperledger BESU in a permissioned setup using IBFT 2.0 consensus. Hyperledger BESU uses internally an embedded RocksDB instance for storing linked blocks (the journal of transaction) and world state (the ledger). Hyperledger BESU can instantiate and execute smart contracts for supporting the use cases of i3-MARKET framework (Figure 5.1).

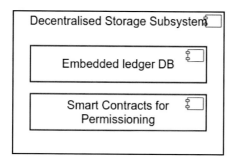

Figure 5.1 Decentralized storage subsystem.

The components depending on the decentralized storage subsystem uses Hyperledger BESU's native JSON-RPC-based interface. A separate interface layer for accessing (or limiting access to) decentralized storage is not planned, as the nodes of the decentralized storage will already validate all transactions submitted to the ledger. The diagram of a distributed storage subsystem is shown in Figure 5.2. The subsystem consists of database nodes. The database provides an SQL interface to other i3-MARKET framework components. The software implementation database is CockroachDB that can be accessed via PostgreSQL-compatible wire protocol for which a large number of client libraries exist in different languages and platforms. Only secure access (TLS with mutual authentication) to the database will be enabled; hence, all clients need to use private keys and valid certificates to access the database.

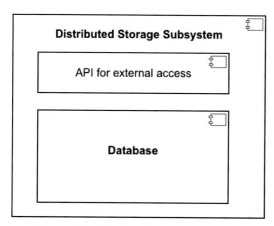

Figure 5.2 Distributed storage subsystem.

The distributed storage component is an internal component with no external access. That is to say that it will have connections only with other trusted services within the i3-MARKET Backplane. Even though this simplifies the necessary measures in terms of authentication and authorization, it is still needed to secure machine-to-machine connections between the i3-MARKET services since they can be deployed on shared infrastructure.

The authentication and authorization solution relies on providing the distributed storage behind a TLS server endpoint and requiring TLS client certificates for the different connecting services. The setup guarantees end-to-end security between the distributed storage service and any of its client services.

The governance of the certificates has followed up to now the *keep it simple* approach. The distributed storage system is in charge of issuing the servers' and clients' certificates, unless the instance has its own certificate authority (CA), in which case the CA is responsible for issuing server certificates to the distributed storage server component and client certificates to the clients.

5.2.1 Service availability

The storage subsystem is a critical component of the i3-MARKET network contributing to the proper functioning of the platform. Hence, appropriate measures in the form of design, choice of technologies, and deployment have to be applied. Fortunately, the two main subsystems used in the storage solution already have strong built-in availability features that are summarized below.

5.2.1.1 Distributed storage

The distributed storage solution is based on a CockroachDB server. Initially, the database was deployed as a global cluster of database nodes; however, after the initial testing of the entire network, a couple of issues were discovered. First, the deployment of a CockroachDB instance and connecting it to a cluster is not an automated process, but rather manual as configuration steps must be tightly coordinated between the nodes. This contradicts with the overall concept of i3-MARKET, which should be operable without any central administration.

The second and far greater problem, which was eventually acknowledged, is that each node in a CockroachDB cluster has equal rights with full administrative privileges over the cluster. This is a problem because any node can alter data and there is no consensus mechanism to agree on the changes. Furthermore, in case of the rise of a rogue node could potentially lead to full erasure or silent corruption of the entire database.

Therefore, a decision was made to replace the global cluster with independent clusters deployed at each i3-MARKET instance. In this deployment mode, each instance is responsible for its own operation and a configuration mistake in one instance, or a malicious act cannot affect the stored data at other instances. There was only one implication to this change – SEED Index would not work in such a setup anymore. As a result, the index was migrated from the distributed storage to the decentralized storage.

5.2.1.2 Decentralized storage

The decentralized storage used in the platform is a Hyperledger BESU network, which uses the IBFT 2.0 (proof of authority) consensus protocol. In this network, there are four validator nodes based on the genesis configuration stored in the corporative Nexus. In this configuration, there are three accounts to be used by the i3-MARKET federation.

The federated search engine index service uses the Hyperledger BESU blockchain as its storage backend. For this purpose, a smart contract storing the endpoints of all SEED instances along with the associated data categories has been deployed on the blockchain.

In this scenario, different components like auditable accounting, SEED Index, etc., are capable to deploy and manage smart contracts and transactions over those accounts.

5.2.2 Verifiable database integrity

The purpose of the VDI is to provide an infrastructure for data to be stored in a way that its presence, or lack thereof, can be cryptographically proven. It takes advantage of the blockchain technology to determine the integrity of the data it contains.

The VDI component consists of a library that implements a Compact Sparse Merkle Tree (CSMT)[1] and exposes an API that allows for data to be inserted, retrieved, updated, and removed. The API consumer can later obtain proofs of membership/non-membership and verify those proofs against the existing Merkle tree.

---

[1] Compact Sparse Merkle Trees: https://eprint.iacr.org/2018/955.pdf

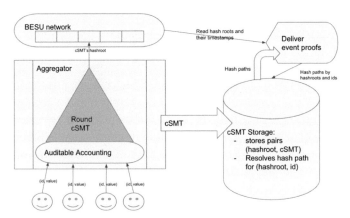

Figure 5.3 VDI integration with auditable accounting.

This data structure works on the same principles of verifiable maps[2], periodically generating a root hash that, after been made public, can guarantee the integrity of the data at that point in time. Membership or non-membership of any given key can be cryptographically proven against the Merkle tree.

• **Integration:**

The VDI is not a separate application on its own but is integrated into the auditable accounting component, as can be seen in Figure 5.3. The MerkleTreeService class exposes methods to create a CSMT tree from the array of hashes obtained from registries. The individual proofs are then stored along with their corresponding root hash into the registries repository in the database. The unregistered blockchain record is also stored in the database along with the serialized Merkle tree data itself.

The library[3] is implemented in TypeScript and distributed as a node package. The main data structure of the VDI is a class called CSMT. This class keeps a map of all the nodes that are stored in the tree as well as the tree's root hash. On an empty tree, the root hash is initialized as a zero node. Data (in key-value format) can then be inserted into the tree, producing new nodes and updating the root hash.

What follows is a summary of the functions that are available for the consumers of the CSMT class.

---

[2] Verifiable data structures, p. 2: https://continusec.com/static/VerifiableDataStructures.pdf
[3] https://github.com/i3-MARKET-V3-Public-Repository/SP4-VerifiableDatabaseIntegrity

- **Add:**

This method adds a given key-value pair to the tree. The key has to be in byte array format. The value can be any arbitrary string and is optional. If the provided key already exists in the tree, an error is returned, and no duplicate keys are allowed. The data is combined in a new array that contains the key in a hexadecimal format, the hash of the value and an entry mark that flags this as a leaf node. This data is then inserted into the nodes map, where the hash of the data is, in itself, the key for this record in the map. Finally, the tree's root hash is recalculated.

- **Insert:**

This is a convenient method to insert data in bulk to the tree. This method just validates the data and calls the method *add* above on each individual element.

- **Get:**

This method looks up for a given key in the nodes map. It returns the hash of the corresponding value for the key, or *undefined* if the key is not present.

- **Delete:**

This method looks up for a given key in the nodes map and removes it. The tree's root hash is then recalculated based on the remaining nodes. If no key is found, an error is returned.

- **Create proof:**

This method looks up for a given key in the nodes map and creates a new proof object. The proof object contains the data itself (if present), the chain of additional nodes along the tree traversal, the root hash of the tree, and a membership flag (true if the given key is present in the tree, false otherwise).

- **Verify proof:**

When the consumer has a proof object, it can verify whether that proof matches against the existing tree by calling this method. It verifies whether the root hashes and node chain (in case of membership) matches with what the CSMT class has stored internally. It returns true if the proof matches. If it returns false, it means that either the proof does not belong to this tree or that the proof was tampered with.

5.2.3 Federated query engine index management

The decentralized storage sub-component of data storage provides function-ality to manage an index used for semantic data discovery. The federated query engine index supports federated queries, a concept implemented by the semantic engine. The distributed storage plays a vital role in supporting the verifiable data integrity, non-repudiation service, and auditable accounting.

Decentralized storage implements two main use cases – managing the index and querying the index – in order to provide the required functionality to the semantic engine for accessing the content of the index.

The index is a collection of data categories together with the endpoint location addresses of the corresponding semantic engines. One semantic engine is not limited to storing offerings belonging to one category but to several of them. Hence, the index contains one to many relationships, linking a specific semantic engine to a set of data categories.

Each semantic engine instance has a private key, while the corresponding public key serves as an identifier that is associated with a set of SEED Index records. The private key is needed to update corresponding index records. Moreover, in order to pay for update transaction, the SEED account must have enough resources. The owner of the SeedsIndexStorage smart contract can assign administrator roles to other keys that can update records stored under any public key.

Every new marketplace joining the i3-MARKET network will connect to the decentralized storage through a semantic engine. If the marketplace has been around for a while, the marketplace has most probably stored offerings metadata. This metadata should also be stored in the SEED to participate in federated queries. Therefore, such a marketplace would have to populate the index by inserting category information to the decentralized storage.

- **Manage:**

In order to provide the most recent and accurate information to the semantic engines in the i3-MARKET network, the index must be kept up to date at all times. Therefore, functions – insert, update, and delete – to maintain the index are required. All these activities are limited to registered i3-MARKET nodes only and the authentication uses self-signed certificates.

- **Insert:**

Before an i3-MARKET instance receives any data offering registrations, the semantic engine has no reason to insert any content into the index. Although it is possible to insert an empty entry containing the endpoint address and an empty category list to the index, it is recommended to keep the index clean of unnecessary information. After receiving the first data offering, the semantic engine inserts the first entry to the index, revealing to other i3-MARKET instances the category of offerings stored in that specific semantic engine.

• **Update:**

Over the course of the market lifecycle, data offerings of different data categories are stored in a single marketplace. Upon the registration of a data offering belonging to a category that is not yet present in the semantic engine, the semantic engine updates the index with relevant information (data category, endpoint address, etc.) by inserting a new entry to the database.

• **Delete:**

The final management activity of the index lifecycle allows the removal of entries from the index. It is the responsibility of the semantic engine to keep the index up to date; therefore, redundant and outdated information is removed from the index. In the event of closing down of an i3-MARKET marketplace instance, either temporarily for maintenance or indefinitely, the semantic engine has to remove unavailable content from the index. Moreover, this function should be accessible by a system administrator to remove relevant entries from the index, in case of a sudden shut down of a marketplace/i3-MARKET node.

• **Query:**

In case a semantic engine needs to perform a federated query among all other instances in the i3-MARKET network, the index shall provide input to the federated query. The semantic engine firstly queries the index with relevant parameters (data category, description, etc.) and the distributed storage shall return information from the index indicating which i3-MARKET instance contains the data that the SEED is looking for.

5.3 Diagrams

Federated query engine index management:

The sequence diagram in Figure 5.4 shows the interaction of the decentralized storage on the SEED regarding the federated query engine index management. Each function – insert, update, delete, and query – has been depicted on a single sequence diagram, as there is no relevant complexity to be shown for each interaction. Index record identifiers (*uuid* in the figure) are derived from node public keys via cryptographic hashing and all requests must be signed with an authorized key (e.g., corresponding private key).

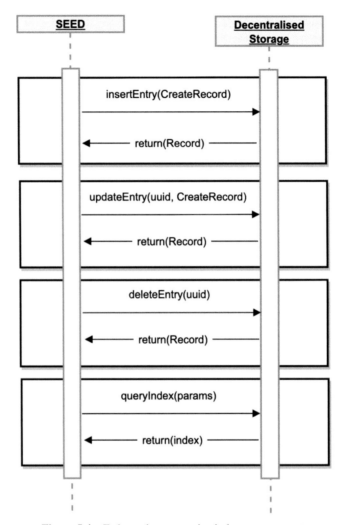

Figure 5.4   Federated query engine index management.

Verifiable data integrity:

The sequence diagram in Figure 5.5 demonstrates the integration of verifiable data integrity with the auditable accounting subsystem. All features are displayed on a single diagram, as there is no specific complexity within the functions.

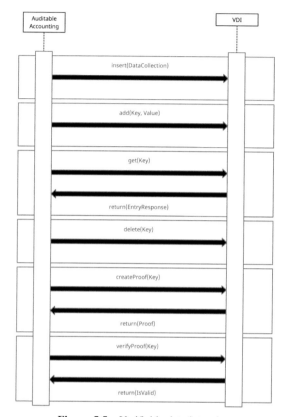

Figure 5.5 Verifiable data integrity.

5.4 Interfaces

The distributed storage subsystem does not expose a bespoke API for internal or external services. Each system within the Backplane uses the storage component's out-of-box means for connectivity.

Likewise, the API provided by the decentralized storage comes out-of-box with the solution. The service can be accessed via JSON-RPC protocol offered by Hyperledger BESU. Please refer to *BESU Documentation*[4] for the details of the API provided by Hyperledger BESU and client libraries.

The SEEDS Index solution consists of a Java library, called SeedsIn-dex, which provides wrappers for the smart contract and utility functions for

---

[4] Hyperledger BESU documentation, https://besu.hyperledger.org/en/stable/

convenience. The SeedsIndex library uses Web3j[5] library for accessing the BESU network. The library interface is documented by extensive Javadoc comments and a complete usage example included in the library.

5.5 Background Technologies

The decisions for the choice of selected technologies in order to satisfy the high-level capabilities are:

- Hyperledger BESU to satisfy decentralized storage.
- CockroachDB[6] to satisfy storage requirements.

Hyperledger BESU is an open-source Ethereum client. The decision to select Hyperledger BESU to satisfy the needs for decentralized storage has been made based on the following assumptions:

- Self-sovereign identity and access management has decided to base the reference implementation on Veramo, which specifically requires Ethereum-based blockchains.
- Auditable accounting and data monetization require smart contract whose functionalities are easily satisfied on Ethereum-based blockchains.

CockroachDB is a relational database management system. CockroachDB has been chosen as the storage solution due to previous experience of the technology by partners. Moreover, it is highly scalable, designed to deliver fast access and resilient to network outages. The only shortcoming of the chosen technology is the lack of features guaranteeing data integrity in case of the presence of malicious (e.g., because of honest mistakes or sophisticated external attacks) users. As all nodes of a CockroachDB cluster that is a part of an i3-MARKET instance are under the control of that instance, this shortcoming is not relevant in the i3-MARKET architecture. For client authentication in CockroachDB, mutual TLS authentication is used.

The semantic search engine index is implemented as a smart contract (written in Solidity programming language), which is deployed on BESU blockchain for distributed access. Updates to the index are authorized with digital signatures.

---

[5] https://www.web3labs.com/web3j-sdk
[6] CockroachDB, https://www.cockroachlabs.com/product/

6

Data Access and Transfer – Design System Principles

6.1 Objectives

The i3-MARKET Data Access & Transfer is a component that defines a secured data access and transfer mechanism allowing an encrypted path between data providers and consumers.

The data access API is the interface via which data consumers gain access to the data offered by a data provider or data space. Since this open interface enables direct interactions among stakeholders of different data spaces/marketplaces, we need not only an open interface specification that can be implemented by all but also a high level of security, as the data exchange might involve sensitive data, e.g., personal data or commercial data.

Since a data exchange shall be only authorized once all involved stakeholders, i.e., data owner, data provider, and data consumer, have signed a smart contract, the data access API must be securely linked with and controlled by the i3-MARKET Backplane. Moreover, for the monetization of data assets based on the crypto currency, the i3-MARKET Backplane must be reliably informed about the quantities of the exchanged assets. This is especially a challenging task due to the decentralized architecture (i.e., the direct, peer-to-peer access interface between data providers and consumers).

Authentication, authorization, and data transfer are the core features of the data access API. Authentication is performed by the i3-MARKET identity provider. The user is authenticated using verifiable claims. After successful authentication, an access token is issued, which contains the user role (data consumer and data provider). If a data consumer tries to access the data provider without a valid access token, it will be redirected to the i3-MARKET identity provider. A data provider validates the access token using a service provided by the Backplane. The data transfer takes place using the non-repudiation protocol. A binary data transfer service based on the non-repudiation protocol was implemented. The service offers support for concurrent data transfer and activity logging integrated with the data transparency subsystem.

The innovative elements of data access API are the following:

1. **Integration of the non-repudiable protocol for secure data transfer:** The user authentication is realized by providing the Verifiable Credentials issued by the i3-MARKET identity provider. An access token is retrieved, and the consumer is authorized for data transfer, while the dataset is split into fixed size blocks transferred one by one. The security of the transfer is enforced by an encryption mechanism implemented with symmetric keys, unique for each data block.
2. **Integration with the i3-MARKET Backplane for data transfer monitoring:** Data transfer tracking and monitoring component measures the amount of transferred data and logs this information, which is transferred to the i3-MARKET Backplane.
3. **Integration with the i3-MARKET smart contract:** The data parameters and characteristics are retrieved by querying the smart contract.

6.2 Technical Requirements

For data access API, the capabilities described below have been defined. They are structured as epics and have been documented in a Trello board as shown in Tables 6.1– 6.10.

Table 6.1   Authentication and authorization – epics.

Name	Description	Labels
Policy management	Policy is a set of rules that define how to protect the assets in order to provide trust, security, and privacy. Policy management component is in charge of enforcing the rule set provided by i3-MARKET Backplane inside of the data access system.	Epic
Role management	A role is a set of policies attached to an entity in order to define the access that entity has within the i3-MARKET data access system. The role management component is in charge with fetching the list of policies and verifying them against the data access system.	Epic

Table 6.2 Authentication and authorization – user stories.

Name	Description	Labels
Intercept access attempts	As a data provider, I want to intercept the data access API access attempts so that I can check the policy	User Story
Check attempt against rule set	As a data provider, I want to check the access attempt of data access API against policy so that I be able to grant access	User Story
Grant access to permitted assets	As a data provider, I want to grant access to assets so that the user can access the data	User Story
Get the list of policies associated with role	As a data provider, I want to access Backplane so that I obtain the list of policies associated with the user's role	User Story
Verify role access	As a data provider, I want to invoke policy management so that I will verify the role access of the user	User Story
Allow or deny access	As a data provider, I want to allow or deny access so that the data can be accessed according to policy	User Story

Table 6.3 Data transfer transparency – epics.

Name	Description	Labels
Data transfer management	Data transfer management is a component that is in charge with the control of the connection between the provider and consumer	Epic
Data transfer tracking	The data transfer tracking component measures the volume of data transferred between the producer and consumer	Epic
Data transfer monitor	The data transfer monitor component communicates with the Backplane before and after the data transfer	Epic

Table 6.4 Data transfer transparency – user stories.

Name	Description	Labels
Initialize the connection	As a data provider, I want to initialize a connection so that I will be able to start the transfer	User Story
Resume the connection	As a data provider, I want to resume the connection so that I will be able to continue the transfer	User Story
Finalize the connection	As a data provider, I want to finalize the connection so that I can conclude the transfer	User Story
Measure transferred data	As a data provider, I want to measure the transferred data so that I can report the information to the Backplane	User Story
Inform i3-MARKET Backplane	As a data provider, I want to inform the Backplane so that the system can track the volume of transferred data	User Story
Invoke linked smart contract	As a data provider, I want to invoke the smart contract so that the data can be transferred according to contractual parameters	User Story

Table 6.5 secure data transfer & anonymization – epics.

Name	Description	Labels
Data encryption	The data encryption component is responsible for the end-to-end process of encoding and decoding of data during transfer between the producer and consumer	Epic
Proxy	The proxy component can be used when the data producer identity needs to be hidden	Epic

Table 6.6 Secure data transfer and anonymization – user stories.

Name	Description	Labels
Key generation and exchange	As a data provider, I want to obtain the encryption key so that I will be able to transfer the data securely	User Story
Transfer encrypted data	As a data provider, I want to transfer encrypted data so that I will be able to enforce the transfer safety and confidentiality	User Story
Decrypt data	As a data consumer, I want to decrypt the transferred data so that I access the transferred data	User Story
Activate proxy	As a data provider, I want to activate the proxy so that I can hide my identity	User Story
Transfer data through proxy	As a data provider, I want to transfer the data through proxy so that my identity remains confidential	User Story

Table 6.7 Data management – epics.

Name	Description	Labels
Batch data transfer management	Batch data transfer management refers to one time data transfer and retrieving one chunk of data in a session	**Epic**
Data stream management	Data stream management component is responsible for the continuous transfer of data based on a subscription, e.g., publish/subscribe mechanism	**Epic**

Table 6.8 Data management – user stories.

Name	Description	Labels
Request batch data	As a data consumer, I want to request a batch of data so that I will be able to obtain the data from a provider	**User Story**
Transfer batch data	As a data provider, I want to transfer a batch data so that I will send the data to consumer	**User Story**
Subscribe to channel	As a data consumer, I want to subscribe to a channel so that I access the streaming data	**User Story**
Trigger data transfer	As a data provider, I want to trigger the data transfer so that the data is sent on a stream	**User Story**
Get data	As a data consumer, I want to get the data so that I can save data locally	**User Story**
Unsubscribe from channel	As a data consumer, I want to unsubscribe from a channel so that I disconnect from the stream of data	**User Story**

Table 6.9 Data access SDK – epics.

Name	Description	Labels
Batch data transfer management	Authentication and authorization are required for users who call the data access API from data access SDK	**Epic**
Data stream management	Data transfer is a component that is responsible for the management of the request data and response	**Epic**

Table 6.10 Data access SDK – user stories.

Name	Description	Labels
Authenticate and authorize the data consumer	As a software developer, I want to authenticate and authorize the consumer so that I will be able to obtain the data from a provider	User Story
Request data	As a software developer, I want to implement a data request so that I get access to data	User Story
Get data	As a software developer, I want to implement the get data so that I can transfer the data locally	User Story

6.3 Solution Design/Blocks

The secure Data Access & Transfer enables data providers to secure registration to access and/or exchange data in a peer-to-peer fashion once the contracts and security mechanisms for identity management have been executed and confirmed. This improves scalability and avoids the need that data providers have to share their data assets with intermediaries (e.g., a marketplace operator). In addition, anonymization can be used to hide the provider's identity.

Data Access & Transfer consists of the following main parts:

- Authentication and authorization
- Policy management
- Role management
- Secure data transfer and anonymization
- Data transfer based on the non-repudiation protocol with support for concurrent threads and logging.

Authentication and authorization:

Authentication: Verifies the identity of the user against the i3-MARKET Backplane.

Authorization: Verifies the permissions the authenticated user has in the i3-MARKET platform allowing to perform authorized actions and granting access to resources.

The authentication and authorization subsystem has the following subcomponents:

Policy management:

Policy is a set of rules that defines how to protect the assets to provide trust, security, and privacy. The policy management component oversees enforcing the rule set provided by i3-MARKET Backplane within the data access system. The responsibilities of the policy management module are:

- Intercept access attempts
- Check attempt against rule set
- Grant access to permitted assets

Role management:

A role is a set of policies attached to an entity in order to define the access that entity has within the i3-MARKET data access system. The role management component is in charge of fetching the list of policies and verifying them against the data access system. The responsibilities of the role management module are:

- Get the list of policies associated with role from Backplane
- verify role access by invoking policy management
- Allow or deny functionalities

Secure data transfer and anonymization:

Secure data transfer and anonymization subsystem has the following components:

Data encryption:

The responsibilities of the data encryption module are:

- Key generation and exchange
- Transfer data in an encrypted way between endpoints
- Decrypt data on the consumer side

Proxy:

The proxy needs to be used when the identity of the data provider needs to be hidden. This feature is optional; therefore, there is no need to implement it if there is no specific requirement referring to the anonymity of the data provider. The responsibilities of the proxy module are:

- Activate the proxy
- Configure the parameters to hide the identity
- Data transfer goes through the proxy

Data transfer transparency:
Data transfer transparency subsystem has the following components:

Data transfer management:
This component is responsible for the management of the connection between provider and consumer and implements the following functionalities:

- Initialize the connection
- Resume the connection
- Finalize the connection

Data transfer tracking:
This component implements the following operation:
- Measure the amount of transferred data.

Data transfer monitor:
The information about how much data was transferred, when the data transfer was initiated and when it was completed, is monitored and the following operations are triggered:

- Inform the i3-MARKET Backplane that the data transfer was performed and report how much data was transferred
- Invoke the linked smart contract

Data management:
Two methods for data transfer are supported by data access API, which are supported by the following modules:

VDI:
One-time data transfer for one chunk of data in a session with the following methods:

- Request data
- Transfer data

Data stream management:
Continuous transfer of data based on a subscription, e.g., publish/subscribe mechanism:

- Subscribe to an offering
- Trigger data transfer – on the producer side
- Get data – on the consumer side
- Unsubscribe

6.4 Diagrams

The process view perspective is presented in the sequence diagrams in Figures 6.1, 6.2, 6.3, and 6.4.

The sequence diagrams of the subsystems listed below are detailed here:

- Authentication and authorization
- Data transfer transparency
- Data management
- Secure data transfer and anonymization

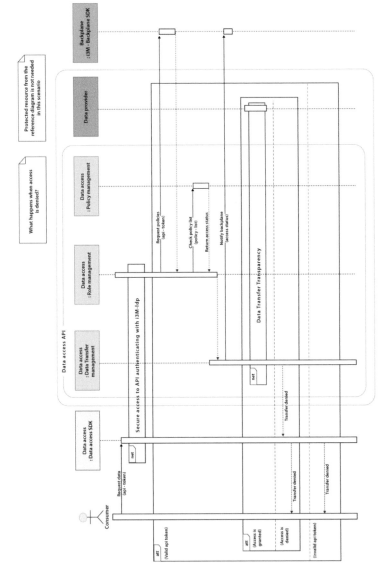

Figure 6.1   Authentication and authorization.

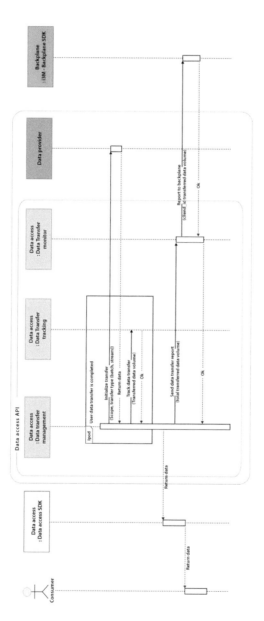

Figure 6.2 Data transfer transparency.

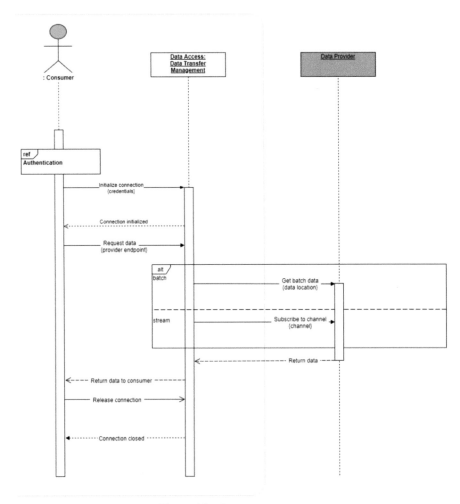

Figure 6.3   Data management.

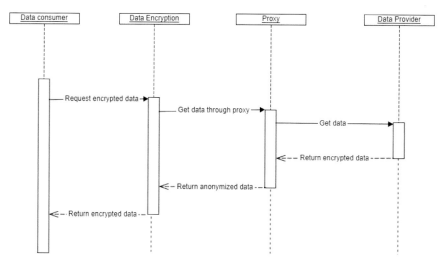

Figure 6.4 Secure.

7

Open-source Strategy

The i3-MARKET consortium is committed to contributing to a reference implementation (community release) of the individual building blocks as well as the overall i3-MARKET data market frameworks to the developer community through an open-source project as shown in Figure 7.1.

The open-source management project structure has been updated to reach the developer and entrepreneur (SMEs) communities largely and facilitate their onboarding or innovation processes. i3-MARKET followed an open-source path using two of the most well-known and established open-source organizations that provide open-source projects hosting: Gitlab and GitHub. We have studied the options to have better impact and acceptance in the developer and SME communities and adopted the procedure and roles for the users of our i3-MARKET open-source project in a way that suited best to the i3-MARKET case.

The i3-MARKET project governance process defines a support and evaluation process to include software improvements as follows:

- **Request for changes or updates:** A technical board identifies any change requests previous to a *major* release, which should be integrated into this *major* release. Before a release, all changes have to be tested by using a pre-production/staging approach.
- **The evaluation of any type of technical request:** A technical board approves a software component or initiates a project in i3-MARKET OSS.
- **The communication of the results from technical experts:** A tagging release strategy is used in order to indicate the impact of the changes made on the i3-MARKET ecosystem.
- **Evaluation of contributions for new commits:** A technical board assesses and evaluates the contributions including documentation in i3-MARKET OSS.

- **Reports and changes report:** A technical board issues a short report, explaining the rationale of the acceptance or the rejection in exceptional cases.

The i3-MARKET team aims to facilitate and simplify development of data services based on i3-MARKET Backplane; any developer should be capable of implementing and developing data services based on i3-MARKET Backplane tools. The i3-MARKET open-source team provides the slack tool (i3-MARKET.slack.com) for a direct communication and conversations with the developers team; the slack channel is used as a direct communication channel and it is open to any developer that is part of the i3-MARKET community but also for those external that want to start engaging with the project.

Developers require technical information that goes beyond high-level descriptions in a website or that a normal software project documentation can provide. The i3-MARKET project has set up an open-source developers portal as an online tool to facilitate the members of the ecosystem to get access to the materials, documentation, technical information, developers know-how, and code. The online tool of the i3-MARKET project is deployed to actively facilitate reaching out not only to the open-source community but also SMEs and entrepreneurs in order to facilitate an easy adoption and building an ecosystem around the i3-MARKET project.

The documentation and specifications are released using the open-source website portal at http://www.open-source.i3-MARKET.eu. Videos showing the progress and use of the developed software tools can be accessed via the i3-MARKET YouTube channel. The community of open-source developers SMEs and entrepreneurs can now easily find instructions that are available at the i3-MARKET open-source portal. This is a live portal, which is a continuous update according to the latest development of the project. The main purpose of releasing this developer-centric portal is to actively enable a channel for reaching out to the open-source community and to allow SMEs and entrepreneurs to get all the latest developments and also download and use the different i3-MARKET available software updates. More specific technical documentations about the components and systems are also available in a specific "Developer Portal" at https://i3-MARKET.gitlab.io/code/Backplane/Backplane-api-gateway/Backplane-api-specification/index.html.

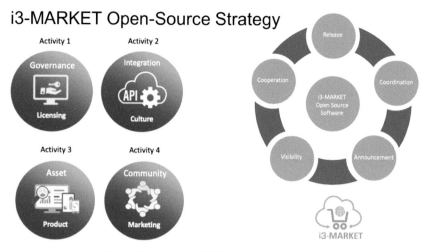

Figure 7.1   i3-MARKET open-source strategy.

The code is available open-source via the establishment of the i3-MARKET spaces on Gitlab, available at https://gitlab.com/i3-MARKET-V3-public-repository/ and GitHub, available at https://github.com/i3-MARKET-V3-public-repository. The i3-MARKET's developers team has made an extra effort to release all the software created in these two well-known platforms as they are among the largest and most popular open-source communities; i3-MARKET has conducted all the necessary efforts to establish an automatic and transparent synchronization mechanism and the OSS governance methodology to support members of both communities. The i3-MARKET Git repository provides the collaborative space and software tools used by the i3-MARKET community. The i3-MARKET Backplane software is included in the current public repositories and further developments will be supported by members and communities.

8

Conclusions

The i3-MARKET project addresses the challenge of being integrative following design methods used in industry and OSS implementation best practices, interoperable by using semantic models that define a common conceptual framework and information model that enables cross-domain data exchange and sharing, and intelligent from the perspective of smart contracts generated automatically and associating those financial operations into a set of software tools that facilitate that data assets can be commercialized via intra-domain or cross-domain almost transparently in a secure and protected digital market environment.

In this book series is presented an overview of the i3-MARKET methodologies and solutions that are the foundations of its software results in the form of a Backplane with a set of software support tools and as a solution addressing the challenge of enabling the coexistence of data spaces with marketplaces for enlarging the European digital market ecosystem.

The i3-MARKET project has built a blueprint open-source software architecture called "i3-MARKET Backplane" that addresses the growing demand for connecting multiple data spaces and marketplaces in a secure and federated manner. The i3-MARKET consortium is contributing with the developed software tools to build the European data market economy by innovating marketplace platforms, demonstrating with three industrial reference implementations (pilots) that a decentralized data economy and more fair growth is possible.

The i3-MARKET architecture design provides adequate and in-house developed building blocks for trustworthy (secure and reliable) data-sharing and exchange of existing data assets for current and new future market-place platforms, with special attention on commercializing data assets from individuals, SMEs, or large industrial corporations. We used and developed the i3-MARKET backplane using open-source technologies that impulse the adoption and exploit the open-source culture, a tendency that, for more than a decade, is hitting the industry markets and that today more and more industries are following.

References

[1] " https://en.wikipedia.org/wiki/System_context_diagram," [Online].

[2] P. Kruchten, "Architectural Blueprints — The "4+1" View Model of Software Architecture," IEEE Software 12, November 1995, pp. 42-50.

[3] J. R. a. I. J. G. Booch, UML User Guide, Addison-Wesley Longman, 1998.

[4] "https://leanpub.com/arc42inpractice/read," [Online].

[5] i3-MARKET, "i3M-Wallet monorepo," [Online]. Available: https://github.com/i3-Market-V3-Public-Repository/SP3-SCGBSSW-I3mWalletMonorepo.

[6] Consensys, "MetaMask," [Online]. Available: https://metamask.io/.

[7] "Trust Wallet," [Online]. Available: https://trustwallet.com/.

[8] Exodus, "Exodus Bitcoin & Crypto Wallet," [Online]. Available: https://www.exodus.com/.

[9] T. Voegtlin, "Electrum Bitcoin Wallet," [Online]. Available: https://electrum.org/.

[10] Validated ID, "VIDChain," [Online]. Available: https://www.validatedid.com/vidchain.

[11] Evernym, "Connect.Me Wallet," [Online]. Available: https://www.connect.me/.

[12] IdRamp, "IdRamp," [Online]. Available: https://idramp.com/.

[13] trinsic, "Identity Wallets," [Online]. Available: https://trinsic.id/identity-wallets/.

[14] ConsenSys, "uPort," [Online]. Available: https://www.uport.me/.

[15] "Twala," [Online]. Available: https://www.twala.io/.

[16] ConsenSys, "Serto," [Online]. Available: https://www.serto.id/.

[17] "Veramo - A JavaScript Framework for Verifiable Data | Performant and modular APIs for Verifiable Data and SSI," [Online]. Available: https://veramo.io/.

[18] "OpenTimeStamps, a timestamping proof standard," [Online]. Available: https://opentimestamps.org/.

[19] Y. Du, H. Duan, A. Zhou, C. Wang, M. H. Au and Q. Wang, "Enabling Secure and Efficient Decentralized Storage Auditing with Blockchain," IEEE Transactions on Dependable and Secure Computing, 2021.

[20] Y. Du, H. Duan, A. Zhou, C. Wang, M. H. Au and Q. Wang, "Towards Privacy-assured and Lightweight On-chain Auditing of Decentralized Storage," 2020 IEEE 40th International Conference on Distributed, pp. 201-211, 2020.

[21] H. Yu and Z. Yang, "Decentralized and Smart Public Auditing for Cloud Storage," IEEE 9th International Conference on Software, pp. 491-494, 2018.

[22] J. Shu, X. Zou, X. Jia, W. Zhang and R. Xie, "Blockchain-Based Decentralized Public Auditing for Cloud Storage," IEEE Transactions on Cloud Computing, 2021.

[23] K. Liu, H. Desai, L. Kagal and M. Kantarcioglu, "Enforceable Data Sharing Agreements Using Smart Contracts," 27 04 2018. [Online]. Available: https://arxiv.org/abs/1804.10645.

[24] E. J. Scheid, B. B. Rodrigues, L. Z. Granville and B. Stiller, "Enabling Dynamic SLA Compensation Using Blockchain-based Smart Contracts," in IFIP/IEEE Symposium on Integrated Network and Service Management (IM), 2019.

[25] Ocean Protocol Foundation with BigchainDB GmbH and Newton Circus (DEX Pte. Ltd.), "Ocean Protocol: A Decentralized Substrate for AI Data and Services," 2019.

[26] The European Parliament and the Council of the European Union, "General Data Protection Regulation (GDPR). Directive 95/46/EC," 27 04 2016. [Online]. Available: https://gdpr-info.eu/.

[27] K. Jensen and L. M. Kristensen, Coloured Petri nets: modelling and validation of concurrent systems, Springer Science & Business Media, 2009.

[28] Digital Asset Holdings, "Digital Asset Modelling Language (DAML)," [Online]. Available: https://daml.com/.

[29] A. M. Antonopoulos, Mastering Bitcoin: unlocking digital cryptocurrencies, O'Reilly Media, Inc., 2014.

[30] I. Bashir, Mastering blockchain, Packt Publishing Ltd, 2017.

[31] D. Yaga, P. Mell, N. Roby and K. Scarfone, "Blockchain technology overview," arXiv preprint arXiv:1906.11078, 2019.

[32] S. Rouhani and R. Deters, "Security, performance, and applications of smart contracts: A systematic survey," IEEE Access, vol. 7, pp. 50759-50779, 2019.

[33] L. Jing and L. Zhentian, "A survey on security verification of blockchain smart contracts," IEEE Access, vol. 7, pp. 77894-77904, 2019.

[34] G. Wood, "Ethereum: A secure decentralised generalised transaction ledger," Ethereum Project White Paper, vol. 151, no. 2014, pp. 1-32, 2014.

[35] H. Chen, M. Pendleton, L. Njilla and S. Xu, "A survey on ethereum systems security: Vulnerabilities, attacks, and defenses," ACM Computing Surveys (CSUR), vol. 53, no. 3, pp. 1-43, 2020.

[36] "Hyperledger Besu," [Online]. Available: https://github.com/hyperledger/besu.

[37] "Solidity," [Online]. Available: https://solidity-es.readthedocs.io/.

[38] "BIP-39," 2021. [Online]. Available: https://github.com/bitcoin/bips/blob/master/bip-0039.mediawiki.

[39] i3-MARKET, "i3M-Wallet OpenApi Specification," 2022. [Online]. Available: https://github.com/i3-Market-V3-Public-Repository/SP3-SCGBSSW-I3mWalletMonorepo/blob/public/packages/wallet-desktop-openapi/openapi.json.

[40] W3C, "Decentralized Identifiers (DIDs) v1.0. Core architecture, data model, and representations," W3C Recommendation, 19 07 2022. [Online]. Available: https://www.w3.org/TR/did-core/.

[41] W3C, "Verifiable Credentials Data Model v1.1.," W3C Recommendation, 03 03 2022. [Online]. Available: https://www.w3.org/TR/vc-data-model/.

[42] F. Román García and J. Hernández Serrano, "i3M-Wallet Base Wallet," [Online]. Available: https://github.com/i3-Market-V3-Public-Repository/SP3-SCGBSSW-I3mWalletMonorepo/tree/public/packages/base-wallet.

[43] F. Román García and J. Hernández Serrano, "SW Wallet," [Online]. Available: https://github.com/i3-Market-V3-Public-Repository/SP3-SCGBSSW-I3mWalletMonorepo/tree/public/packages/sw-wallet.

[44] F. Román García and J. Hernández Serrano, "BOK Wallet," [Online]. Available: https://github.com/i3-Market-V3-Public-Repository/SP3-SCGBSSW-I3mWalletMonorepo/tree/public/packages/bok-wallet.

[45] F. Román García and J. Hernández Serrano, "Wallet Desktop," [Online]. Available: https://github.com/i3-Market-V3-Public-Repository/SP3-SCGBSSW-I3mWalletMonorepo/tree/public/packages/wallet-desktop.

[46] J. Hernández Serrano and F. Román García, "Server Walllet," [Online]. Available: https://github.com/i3-Market-V3-Public-Repository/SP3-SCGBSSW-I3mWalletMonorepo/tree/public/packages/server-wallet.

[47] J. Hernández Serrano and F. Román García, "Wallet Desktop OpenAPI," [Online]. Available: https://github.com/i3-Market-V3-Public-Repositor y/SP3-SCGBSSW-I3mWalletMonorepo/tree/public/packages/wallet-desktop-openapi.

[48] F. Román García and J. Hernández Serrano, "Wallet Protocol," [Online]. Available: https://github.com/i3-Market-V3-Public-Repository/SP3-SCGBSSW-I3mWalletMonorepo/tree/public/packages/wallet-protocol.

[49] F. Román García and J. Hernández Serrano, "Wallet Protocol API," [Online]. Available: https://github.com/i3-Market-V3-Public-Repos itory/SP3-SCGBSSW-I3mWalletMonorepo/tree/public/packages/walle t-protocol-api.

[50] F. Román García and J. Hernández Serrano, "Wallet Protocol Utils," [Online]. Available: https://github.com/i3-Market-V3-Public-Repositor y/SP3-SCGBSSW-I3mWalletMonorepo/tree/public/packages/wallet-protocol-utils.

[51] IDEMIA, "Video proving the integration of IDEMIA's HW Wallet into the i3-MARKET Wallet Desktop application," 2022. [Online]. Available: https://drive.google.com/file/d/1Ai_eoDIzIHczOjzOMBR4ctV5 kbR05NOE/view?usp=share_link.

[52] Bluetooth SIG - Core Specification Workgroup, "Bluetooth Core Specification v2.1 + EDR: Secure Simple Pairing," 2007.

[53] D. Basin, C. Cremers, J. Dreier, S. Meier, R. Sasse and B. Schmidt, "Tamarin Prover," [Online]. Available: http://tamarin-prover.github.io/.

[54] OpenJS Foundation, "Electron," [Online]. Available: https://www.electr onjs.org/.

[55] Ethers JS, "The Ethers Project," [Online]. Available: https://github.com /ethers-io/ethers.js/.

[56] Veramo, "Veramo - A JavaScript Framework for Verifiable Data," [Online]. Available: https://veramo.io/.

[57] OpenAPI, "OpenAPI Initiative," Linux Foundation, [Online]. Available: https://www.openapis.org/.

[58] "Express OpenAPI Validator," [Online]. Available: https://github.com/c dimascio/express-openapi-validator.

[59] TypeDoc, "TypeDoc," [Online]. Available: https://typedoc.org.

[60] J. Hernández Serrano, "i3-MARKET Non-Repudiation Library," 2022. [Online]. Available: https://github.com/i3-Market-V3-Public-Repositor y/SP3-SCGBSSW-CR-NonRepudiationLibrary.

[61] J. Hernández Serrano, "i3-MARKET Conflict Resolver Service," 2022. [Online]. Available: https://github.com/i3-Market-V3-Public-Repositor y/SP3-SCGBSSW-CR-ConflictResolverService.

[62] J. Hernández Serrano, "API of the i3-MARKET Non-Repudiation Library," i3-MARKET, 2022. [Online]. Available: https://github.com /i3-Market-V3-Public-Repository/SP3-SCGBSSW-CR-NonRepudiat ionLibrary/blob/public/docs/API.md.

[63] Panva, "JOSE," [Online]. Available: https://github.com/panva/jose.

[64] Ajv, "Ajv JSON schema validator," [Online]. Available: https://ajv.js.o rg/.

[65] OpenJS Foundation, "Express JS," [Online]. Available: https://expressj s.com/.

[66] Y. Kovacs, S. Stanhke and J. L. Muñoz, "i3-MARKET Smart Contracts," [Online]. Available: https://github.com/i3-Market-V3-Public-Repositor y/SP3-SCGBSSW-I3mSmartContracts.

[67] Hans van der Veer and Anthony Wiles, "Achieving Technical Interoper- ability - the ETSI Approach," in ETSI, 2008.

[68] Mike Ushold, Christopher Menzel, and Natasha Noy. Semantic Integra- tion & Interoperability Using RDF and OWL. [Online]. https://www.w3 .org/2001/sw/BestPractices/OEP/SemInt/

[69] M. Compton et al., "The SSN ontology of the W3C semantic sensor network incubator group," JWS, 2012.

[70] EUROPA. Publications Office of the EU. EU Vocabularies. Controlled Vocabularies. Authority tables. Frequency. https://publications.europa. eu/en/web/eu-vocabularies/at-dataset/-/resource/dataset/frequency~

[71] EUROPA. Publications Office of the EU. EU Vocabularies. Controlled Vocabularies. Authority tables. File type. https://publications.europa.eu /en/web/eu-vocabularies/at-dataset/-/resource/dataset/file-type~

[72] EUROPA. Publications Office of the EU. EU Vocabularies. Controlled Vocabularies. Authority tables. Language. https://publications.europa. eu/en/web/eu-vocabularies/at-dataset/-/resource/dataset/language/

[73] EUROPA. Publications Office of the EU. EU Vocabularies. Controlled Vocabularies. Authority tables. Corporate body. https://publications.eur opa.eu/en/web/eu-vocabularies/at-dataset/-/resource/dataset/corporate- body/

[74] EUROPA. Publications Office of the EU. EU Vocabularies. Controlled Vocabularies. Authority tables. Continent https://publications.europa.eu /en/web/eu-vocabularies/at-dataset/-/resource/dataset/continent

[75] EUROPA. Publications Office of the EU. EU Vocabularies. Controlled Vocabularies. Authority tables. Country. https://publications.europa.eu /en/web/eu-vocabularies/at-dataset/-/resource/dataset/country

[76] EUROPA. Publications Office of the EU. EU Vocabularies. Controlled Vocabularies. Authority tables. Place.~https://publications.europa.eu/e n/web/eu-vocabularies/at-dataset/-/resource/dataset/place

[77] European Commission. Joinup. Asset Description Metadata Schema (ADMS).~https://joinup.ec.europa.eu/solution/asset-description-metad ata-schema-adms

[78] CI/CD with Ansible Tower and GitHub. Available from: https://keithten zer.com/2019/06/24/ci-cd-with-ansible-tower-and-github/

[79] Red Hat Ansible Tower Monitoring: Using Prometheus + Node Exporter + Grafana. Available from: https://www.ansible.com/blog/red-hat-ansib le-tower-monitoring-using-prometheus-node-exporter-grafana

i3-MARKET

Index

About the Editors

Dr. Martin Serrano is a recognized expert on semantic interoperability for distributed systems due to his scientific contribution(s) to using liked data and semantic formalisms like ontology web language for the Internet of Things and thus store the collected sensor's data in the Cloud. He has also contributed to define the data interplay in edge computing using the linked data paradigm; in those works he has received awards recognizing his scientific contributions and publications. Dr. Serrano has advanced the state of the art on pervasive computing using semantic data modelling and context awareness methods to extend the "autonomics" paradigm for networking systems. He has also contributed to the area of information and knowledge engineering using semantic annotation and ontologies for describing data and services relations in the computing continuum. Dr. Serrano has defined the data continuum and published several articles on data science and Internet of Things science and he is a pioneer and visionary on proposing that semantic technologies applied to policy-based management systems can be used as an approach to produce cognitive applications capable of understanding, service and application events, controlling the pervasive services life cycle. A process called bringing semantics into the box, as published in one of his academic books. He has published 5 academic books and more than 100 peer reviewed articles in IEEE, ACM and Springer conferences and journals.

Dr. Achille Zappa is a Post-Doctoral Researcher at Insight, University of Galway. He received BSC/MSC degree in Biomedical Engineering and PHD in Bioengineering from the University of Genoa (Italy), his Ph.D. project was related to semantic web integration, knowledge engineering and data management of biomedical and genomic data and his research interests include semantic web technologies, semantic data mashup, linked data, big data management, knowledge engineering, big data integration, semantic integration in life sciences and health care, workflow management, IoT semantic interoperability, IoT semantic data and systems integration. Dr. Zappa is the W3C Advisory Committee representative for Insight Centre at University of Galway and member of W3C working groups like the HCLS IG, the Web of Things (WoT) IG and WG, the Spatial Data on the Web WG. He currently work with the main Insight Linked Data and Semantic Web Groups and with the UIoT (Internet of Things, stream processing and intelligent systems unit) Research Unit, addressing collaboration with different units and involvement in various projects where he seeks to develop general-purpose linked data analytics platform(s), which enables (a) flexible and scalable data integration mechanisms and (b) flexible use and reuse of data analytics components such as visualization components and analytics methods. Dr. Zappa has an extensive expertise of applying semantic web technologies and linked data principles in health care and life sciences domains.

Mr. Waheed Ashraf is a Senior Software Engineer with extensive experience in Java programming with Spring Boot and Project Management experience with a strong background on microservices systems design and is an AWS Certified person. Mr. Ashraf is a highly skilled senior software engineer, with 10+ years of project related professional experience in developing and implementing software systems and developing and maintaining enterprise applications working for international companies from USA, Australia and Malaysia. Mr. Ashraf is also proficient in agile software development, scrum and continuous integration (Jenkins), Amazon Web Services (AWS) and back-end RDBMS (using SQL in Databases Like Oracle, DB2, MySQL 4.0 and Microsoft SQL Server). He is currently responsible for the design, development and implementation of a federated authentication and authorization infrastructure (AAI) for federated access to data providers in the context of the Federated Decentralized Trusted Data Marketplace for Embedded Finance FAME Horizon Europe project.

Mr. Edgar Fries is Senior System Architect at Siemens AG, Germany. In his early career he acted as project manager and consultant at SIEMENS AG consulting in the field of engineering with a focus on engineering tools and methods for customers in the plant engineering and product business. Fries is graduated from the Technical computer science in Esslingen University of Applied Sciences.

Iván Martínez is project manager and SW architect at Atos, Spain, and a senior researcher at the ARI department of the company AtoS. He graduated in computer science from Technical University of Madrid and in the past few years he has participated in semantic web, cloud, big data and blockchain related industrial and research projects. He has contributed to national research projects such as PLATA, and other Cloud, HPC and big data related projects, such as KHRESMOI, VELaSCCo, TOREADOR, DataBench and BODYPASS mainly leading in the latter's definition and integration of system architecture.

Mr. Alessandro Amicone is an experience project manager at GFT, Italy leading both public funded and commercial market projects. In the first part of his professional career, he worked mainly in projects focusing on coordinating documents management and business process management systems for the bank and insurance industry. In recent years he has been working on Horizon2020 projects and innovative market projects promoting smart communities and technology for digital transformation for and between companies in the industry sector and research communities. The development of processes and management systems mainly focuses on advancing the state of art using software engineering for blockchain, smart contracts and distributed/self-sovereign identity, ensuring cyber-security solutions.

Dr. Pedro Malo is professor at the Electrotechnical Engineering and Computers Department (DEEC) of the NOVA School of Science and Technology (FCT NOVA), Senior Researcher at UNINOVA research institute and Entrepreneur at UNPARALLEL Innovation research-driven hi-tech SME. He obtained an M.Sc. in Computer Science and holds a Ph.D. in Computer Engineering with research interests in interoperability and integrability of (complex) systems with special emphasis on cyber-physical systems/Internet of Things. Pedro coined novel methods and tools such as the plug'n'play interoperability (PnI) solution for large-scale data interoperability and the NOVAAS (NOVA Asset Administration Shell) that establishes the guidelines and methodology for industry digitization by integrating industrial assets into a Industry 4.0 communication backbone. As an entrepreneur, Pedro initiated the development of the IoT Catalogue that aims to be the whole-earth catalogue of the Internet of Things (IoT) – the one-stop-source for innovations, products, applications, solutions, etc. to help users (developers/integrators/advisors/end-users) to take the most advantage of the IoT for the benefit of society, businesses and individuals. Pedro has

20+ years practice in the management, research and technical coordination/development of RTD and innovation projects in ICT domains especially addressing data technologies, systems' interoperability and integration solutions. Pedro is a recognized Project Manager and S&T Coordinator of European/National RTD and industry projects with skills in the coordination of both co-localized and geographical dispersed work teams operating in multidisciplinary and multicultural environments.

Márcio Mateus is project Manager at Unparallel Innovation, Lda Portugal and a Research engineer holding an M.Sc. in electrotechnical and computer engineering from the Faculty of Science and Technology of the Universidade Nova de Lisboa (FCT NOVA). Márcio is an expert in data interoperability measurement techniques and methodologies for complex heterogeneous environments.

Justina Bieliauskaite is Innovations Director at the European Digital SME Alliance with more than 8 years of project lead and management experience (previously she worked in Lithuanian and Belgian NGOs). Justina Bieliauskaite leads the preparation and implementation of Horizon Europe, Digital Europe Programme, Erasmus+ and other tenders/service contracts for the European Commission. She is experienced in coordinating stakeholder engagement, policy analysis and recommendations, SME training, standardization, and communication activities. Justina is currently the main coordinator of the BlockStand.eu project. Currently, Justina is leading DIGITAL SME's Projects and Standardisation teams, and coordinates the internal WG DIGITALIZATION which covers AI, IoT, cloud computing, blockchain and emerging technologies, as well as coordination among digital innovation hubs. Justina holds a Master's degree in Science (cum laude), focusing on political science and international relations, from the Universities of Leiden and Vilnius. Besides her mother-tongue Lithuanian, Justina speaks English, Italian, Russian and German.

Dr. Marina Cugurra is a lawyer specializing in R&I projects, in particular in legal issues of new technologies and Information Society (e.g. AI, GDPR, data ownership, etc.), with a Ph.D. degree at the "Telematics and Information Society" Ph.D. School at University of Florence. She is also an expert in ethical and societal themes related to ICT research and technological developments. She is serving as independent Ethical Expert at European Commission and European Defense Agency. Consolidated experience in national projects and international and European projects. Scientific collaboration with CNIT (National Inter-University Consortium for Telecommunications) and CNR – ITTIG (Italian National Research Council, Institute of Legal Information Theory and Techniques). Legal Advisor in the R&I Division of multinational companies. She has contributed to the activities of the legal working groups of Eu-wide initiatives (EU Blockchain Observatory Forum) and is Chair of the Ethics, Data Protection and Privacy (EDPP) Task Force of the "Citizen's Control of Personal Data" Initiative within Smart City Marketplace.

Printed in the United States
by Baker & Taylor Publisher Services